John Wiley & Sons

New York Chichester
Brisbane Toronto
Singapore

SOLUTIONS MANUAL TO ACCOMPANY

ELEMENTARY DIFFERENTIAL EQUATIONS AND BOUNDARY VALUE PROBLEMS

FOURTH EDITION

AND

ELEMENTARY DIFFERENTIAL EQUATIONS

FOURTH EDITION

WILLIAM E. BOYCE
RICHARD C. DiPRIMA

CHARLES W. HAINES

PREFACE

This supplement has been prepared for use in conjunction with the fourth editions of ELEMENTARY DIFFERENTIAL EQUATIONS AND BOUNDARY VALUE PROBLEMS and ELEMENTARY DIFFERENTIAL EQUATIONS, both by W.E. Boyce and R.C. DiPrima. The supplement contains a sampling of problems from each section of the text. In most cases the complete details in determining the solutions are given while in the remainder of the problems helpful hints are provided. The problems chosen in each section represent, wherever possible, the variety of applications that are covered in the written material of the text, thereby providing the student a complete set of examples from which to learn.

In order to simplify the text, the following abbreviations have been used:

D.E.	differential equation(s)
O.D.E.	ordinary differential equation(s)
P.D.E.	partial differential equation(s)
I.C.	initial condition(s)
I.V.P.	initial value problem(s)
B.C.	boundary condition(s)
B.V.P.	boundary value problem(s)

I wish to express my appreciation to Mr. Surendra K. Gupta and Ms. Marie J. Haines for their assistance in the preparation of the figures and to Mrs. Susan A. Hickey for her excellent typing and proofreading of all stages of the manuscript.

Charles W. Haines
Departments of Mathematics
  and Mechanical Engineering
Rochester Institute of Technology
Rochester, New York
November, 1985

CONTENTS

CHAPTER 1 1

CHAPTER 2 2

CHAPTER 3 38

CHAPTER 4 77

CHAPTER 5 112

CHAPTER 6 123

CHAPTER 7 140

CHAPTER 8 177

CHAPTER 9 192

CHAPTER 10 219

CHAPTER 11 252

CHAPTER 1

1b. The D.E. is second order since there is a second derivative of y appearing in the equation. The equation is nonlinear due to the y^2 term.

2b. To show that $y_1(x) = e^{-3x}$ is a solution of the D.E. differentiate y_1 twice to find $y_1'(x) = -3e^{-3x}$ and $y_1''(x) = 9e^{-3x}$. Substitution of these into the D.E. yields $9e^{-3x} + 2(-3e^{-3x}) - 3(e^{-3x}) = (9-6-3)e^{-3x} = 0$.

3b. Differentiating e^{rx} twice and substituting into the D.E. yields $r^2 e^{rx} - e^{rx} = (r^2-1)e^{rx}$. If $y = e^{rx}$ is to be a solution of the D.E., then the last quantity must be zero for all x and thus $r^2-1 = 0$ since e^{rx} is never zero.

4a. Differentiating x^r twice and substituting into the D.E. yields $x^2[r(r-1)x^{r-2}] + 4x[rx^{r-1}] + 2x^r = [r^2 + 3r + 2]x^r$. If $y = x^r$ is to be a solution of the D.E., then the last term must be zero for all x and thus $r^2 + 3r + 2 = 0$.

5d. The D.E. is second order since there are second partial derivatives of u(x,y) appearing. The D.E. is nonlinear due to the product of u(x,y) times u_x (or u_y).

6b. Since $\partial u_1/\partial t = -\alpha^2 e^{-\alpha^2 t} \sin x$ and $\partial^2 u_1/\partial x^2 = -e^{-\alpha^2 t}$. $\sin x$ we have $\alpha^2[-e^{-\alpha^2 t} \sin x] = -\alpha^2 e^{-\alpha^2 t} \sin x$, which is true for all t and x.

CHAPTER 2

Section 2.1, Page 20

The method of solving problems 1 through 14 follows the pattern of Example 2, since Example 1 can be considered a special case of Example 2 with $p(x) = a$. The first eight problems do not have I.C., thus the constant of integration (c) cannot be determined. Eq.(21) is used in each case to find the integrating factor $\mu(x)$.

1. $\mu(x) = \exp(\int^x 3dt) = e^{3x}$. Thus $e^{3x}(y'+3y) = e^{3x}(x+e^{-2x})$

    or $\frac{d}{dx}(ye^{3x}) = xe^{3x} + e^x$. Integration of both sides

    yields $ye^{3x} = \frac{1}{3}xe^{3x} - \frac{1}{9}e^{3x} + e^x + c$, and division by

    e^{3x} gives the solution. Note that $\int xe^{3x}dx$ is evaluated

    by integration by parts, with $u = x$ and $dv = e^{3x}dx$.

2. $\mu(x) = e^{-x}$ 3. $\mu(x) = e^x$

4. $\mu(x) = \exp(\int^x \frac{dt}{t}) = e^{\ln x} = x$, so $(xy)' = 3x \cos 2x$, and

    integration by parts yields the solution.

6. The equation must be divided by x so that it is in the

    form of Eq. (4): $y' + (2/x)y = (\sin x)/x$. Thus

    $\mu(x) = \exp(\int^x \frac{2dt}{t}) = x^2$, and $(x^2 y)' = x \sin x$.

    Integration then yields $x^2 y = -x \cos x + \sin x + c$.

7. $\mu(x) = e^{x^2}$ 8. $\mu(x) = \exp(\int^x \frac{4tdt}{1+t^2}) = (1+x^2)^2$

9. $\mu(x) = e^{-x}$ and $y = 2(x-1)e^{2x} + ce^x$. To find the value

    for c, set $x = 0$ in y and equate to 1, the initial value

    of y. Thus $-2+c = 1$ and $c = 3$, which yields the unique

    solution.

11. $\mu(x) = \exp(\int^x \frac{2dt}{t}) = x^2$ and $y = x^2/4 - x/3 + 1/2 + c/x^2$.

Setting $x = 1$ and $y = 1/2$ we have $c = 1/12$.

12. $\mu(x) = x^2$ 13. $\mu(x) = e^{-2x}$

14. $\mu(x) = x^2$. Thus $(x^2 y)' = x \sin x$ and $x^2 y = -x \cos x +$

$\sin x + c$. Setting $x = \pi/2$ and $y = 1$ yields

$c = \pi^2/4 - 1$.

15. The D.E. as given is nonlinear. However, if we think of

y as the independent variable and x as the dependent

variable then the D.E. can be written as $\frac{dx}{dy} = e^y - x$ or

$\frac{dx}{dy} + x = e^y$. The integrating factor is then $\mu(y) = e^y$

and the I.C. is $x(0) = 1$.

16a. To show that $\phi(x) = e^{2x}$ is a solution of the D.E., take

its derivative and substitute into the D.E.

17. $[c\,\phi(x)]' + p(x)\,[c\,\phi(x)] = c\,[\phi'(x) + p(x)\,\phi(x)] = 0$

since $\phi(x)$ satisfies the given D.E.

18. $[y_1(x) + y_2(x)]' + p(x)\,[y_1(x) + y_2(x)] =$

$y_1'(x) + p(x)y_1(x) + y_2'(x) + p(x)y_2(x) = 0 + g(x)$.

19. This problem demonstrates the central idea of the method

of variation of parameters for the simplest case. The

solution (ii) of the homogeneous D.E. is extended to the

corresponding nonhomogeneous D.E. by replacing the

constant A by a function A(x), as shown in (iii).

20a.   Assume $y(x) = A(x) \exp(-\int^x(-2)dt) = A(x)e^{2x}$.

Differentiating $y(x)$ and substituting into the D.E.

yields $A'(x) = x^2$ since the terms involving $A(x)$ add to

zero. Thus $A(x) = x^3/3 + c$, which substituted into $y(x)$

yields the solution.

20b. $y(x) = A(x) \exp(-\int^x \frac{dt}{t}) = A(x)/x$.

Section 2.2, Page 27

Problems 1 through 4 follow the pattern of solution from
section 2.1.

1. $\mu(x) = x$

2.   Remember to divide both sides of the D.E. by x^2 to get

$\mu(x) = \exp\int^x \frac{3}{t} dt = e^{3lnx} = x^3$.

3. $\mu(x) = \exp \int^x \tan t \, dt = \exp \int^x \frac{\sin t}{\cos t} dt = \exp(-\ln \cos x) =$

exp $\ln(1/\cos x) = \sec x$. Multiplying both sides of the

D.E. by sec x and simplifying yields $(y \sec x)' =$

$2x \sin x$. To solve this the right side must be

integrated by parts with $u = x$ and $dv = \sin x \, dx$. Thus

$y \sec x = -2x \cos x + 2 \sin x + c$, which yields the

solution. For all steps we must have $-\pi/2 < x < \pi/2$ for

the functions to be defined.

4. $\mu(x) = \exp \int^x \frac{2}{t} dt = x^2$

In problems 5 through 11, we must determine the largest
interval in which the functions p and g are continuous and
which contain the initial point, and also to determine the
constant c of the general solution. The procedure follows
Example 1 of this section.

5.   Writing the D.E. in the form of Eq.(1) of this section

we have $y' + (2/x) y = x-1 + 1/x$, $y(1) = 1/2$. Thus $p(x)$

and $g(x)$ are continuous on any interval not containing

the origin. Since the initial point is 1, the solution

will be valid on $0 < x < \infty$. $\mu(x) = x^2$ and thus $(x^2 y)' =$

$x^3 - x^2 + x$ and $y = \frac{1}{4}x^2 - \frac{1}{3}x + \frac{1}{2} + \frac{c}{x}$. Substituting x=1

and $y = 1/2$, we obtain c = 1/12, which gives the desired

solution.

7. $p(x) = \cot x$ and $g(x) = 2 \csc x$ are both continuous for

$n\pi < x < (n+1)\pi$, where n is any integer. Since the

initial point is $x = \pi/2$, we choose n = 0 and conclude

that the solution will be valid on $0 < x < \pi$. Now $\mu(x) =$

$\exp \int^{x} \cot t \, dt = \sin x$ and thus $(y \sin x)' = 2$, which

gives the general solution $y = x \csc x + c \csc x$.

Setting $x = \pi/2$ and y = 1 we find $c = 1 - \pi$.

8. $\mu(x) = x^2$

9.   See explanation of interval in solution to problem 7.

10. $p(x) = \frac{2(1+x)}{x(2+x)}$ and $g(x) = \frac{1+3x}{x(2+x)}$. Thus $p(x)$ and $g(x)$ are

continuous on $-\infty < x < -2$, $-2 < x < 0$ and $0 < x < \infty$.

Since we are given $y(-1) = 1$, the solution will be valid

on $-2 < x < 0$. $\mu(x) = x^2 + 2x$.

11. $p(x)$ and $g(x)$ are continuous for all x and thus so is

the solution. $\mu(x) = e^x$.

12. $\mu(x) = x^2$ 13. $\mu(x) = 1/x$

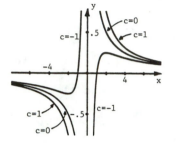

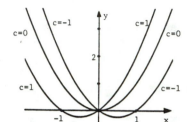

14. $\mu(x) = 1/x$. Solution

exists only for $x \geq 0$

since $g(x) = x^{1/2}$

is defined only for $x > 0$.

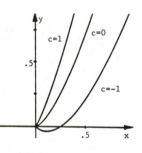

15. $\mu(x) = x$

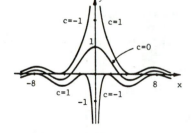

17b.   The limit as $x \to \infty$ of $[\frac{\sqrt{\pi}}{2}$ erf(x) + y(0)] must be zero

for $\lim_{x \to \infty} y$ to be finite. In this case L'Hopital's Rule

applies.

19. $\mu(x) = e^{ax}$ so the D.E. can be written as $(e^{ax}y)' =$
       $be^{ax}e^{-\lambda x} = be^{(a-\lambda)x}$. If $a \neq \lambda$, then integration and
       solution for y yields $y = [b/(a-\lambda)]e^{-\lambda x} + ce^{-ax}$. Then
       $\lim_{x \to \infty} y$ is zero since both λ and a are positive numbers.
       If $a = \lambda$, then the D.E. becomes $(e^{ax}y)' = b$ which yields
       $y = (bx + c)/e^{\lambda x}$ as the solution. Using L'Hopital's
       Rule gives $\lim_{x \to \infty} y = \lim_{x \to \infty} \dfrac{(bx+c)}{e^{\lambda x}} = \lim_{x \to \infty} \dfrac{b}{\lambda e^{\lambda x}} = 0$

20a. $\mu(x) = e^{2x}$. Since g(x) is continuous on the interval
       $0 \le x \le 1$ we may solve the I.V.P. $y_1' + 2y_1 = 1$, $y_1(0) = 0$
       on that interval to obtain $y_1 = 1/2 - (1/2)e^{-2x}$,
       $0 \le x \le 1$. g(x) is also continuous for $1 < x$; and hence we
       may solve $y_2' + 2y_2 = 0$ to obtain $y_2 = ce^{-2x}$, $1 < x$. The
       solution y_1 of the original I.V.P. must be continuous
       (since its derivative must exist) and hence we must
       choose c in y_2 so that y_2 at 1 has the same value as y_1
       at 1. Thus
       $ce^{-2} = 1/2 - e^{-2}/2$ or $c = 1/2 (e^2-1)$ and we obtain

       $y = \begin{cases} 1/2 - 1/2e^{-2x} & 0 \le x \le 1 \\ 1/2(e^2-1)e^{-2x} & 1 \le x \end{cases}$ and $y' = \begin{cases} e^{-2x} & 0 \le x \le 1 \\ (1-e^2)e^{-2x} & 1 < x. \end{cases}$

       Evaluating the two parts of y' at $x_0 = 1$ we see that
       they are different, and hence y' is not continuous at
       $x_0 = 1$.

20b. $\mu(x) = e^{2x}$ for $0 \le x \le 1$ and $\mu(x) = e^x$ for $1 < x$.

21a.   For $n = 0,1$, the D.E. is linear and Eqs. (3) and (4)

       apply.

21b.   Let $v = y^{1-n}$, then $\frac{dv}{dx} = (1-n)y^{-n}\frac{dy}{dx}$ so

       $\frac{dy}{dx} = \frac{1}{1-n}y^n\frac{dv}{dx}$, which makes sense when $n \ne 0,1$.

       Substituting into the D.E. yields $\frac{y^n}{1-n}\frac{dv}{dx} + p(x)y =$

       $q(x)y^n$ or $v' + (1-n)p(x)y^{1-n} = (1-n)\ q(x)$. Setting

       $v = y^{1-n}$ then yields a linear D.E. for v.

22. $n = 3$ so $v = y^{-2}$ and $\frac{dv}{dx} = -2y^{-3}\frac{dy}{dx}$ or $\frac{dy}{dx} = -\frac{1}{2}\ y^3\frac{dv}{dx}$.

       Substituting this into the D.E. gives

       $-\frac{1}{2}\ y^3\frac{dv}{dx} + \frac{2}{x}\ y = \frac{1}{x^2}\ y^3$. Simplifying and setting $y^{-2} = v$

       then gives the linear D.E. $v' - \frac{4}{x}\ v = -\frac{2}{x^2}$, where $\mu(x) =$

       $\frac{1}{x^4}$ and $v(x) = cx^4 + \frac{2}{5x} = \frac{2+5cx^2}{5x}$. Thus $y =$

       $\pm[5x/(2+5cx^2)]^{1/2}$.

23. $n = 2$ so $v = y^{-1}$ and $\frac{dy}{dx} = -y^2\frac{dv}{dx}$. Thus the D.E. becomes

       $-y^2\frac{dv}{dx} - ry = -ky^2$ or $\frac{dv}{dx} + rv = k$. Hence $\mu(x) = e^{rx}$ and

       $v = k/r + ce^{-rx}$. $y = 1/v$ then yields the solution.

24. $n = 3$ 25. $n = 3$

Section 2.3, Page 36

1b.   Theorem 2.2 guarantees a unique solution to the D.E.

through any point (x_0, y_0) such that $x_0{}^2 + y_0{}^2 < 1$ since

$\frac{\partial f}{\partial y} = -y (1-x^2-y^2)^{1/2}$ is defined and continuous only for

$1-x^2-y^2 > 0$. Note also that $f = (1-x^2-y^2)^{1/2}$ is defined

and continuous in this region as well as on the

boundary $x^2+y^2 = 1$. The boundary can't be included in

the final region due to the discontinuity of $\frac{\partial f}{\partial y}$ there.

1c. We are guaranteed a solution to the D.E. passing

through any point in the xy plane since $f = 2xy/(1+y^2)$

and $\frac{\partial f}{\partial y} = 2x(1-y^2)/(1+y^2)^2$ are defined and continous for

all x and y.

3.   If $\phi = [2(x+c)]^{-1/2}$ then $\phi' = -[2(x+c)]^{-3/2}$ and thus

$\phi' + \phi^3 = 0$. Setting $x = 1$ in ϕ yields

$\phi(1) = [2(1+c)]^{-1/2} = 2$ and solving for c gives $c = -7/8$.

The solution $\phi(x) = [2(x-7/8)]^{-1/2}$, which passes through

the initial point $(1,2)$ is valid for all $x > 7/8$ because

the quantity $(x-7/8)^{-1/2}$ is defined only for those values

of x.

4.   Since $\phi' = 1/2[1 + \frac{2}{3} \ln (1+x^3)]^{-1/2} [\frac{2}{3} \frac{3x^2}{1+x}] = \frac{x^2}{1+x^3} \frac{1}{\phi}$

and $\phi(0) = [1 + \frac{2}{3}\ln(1)]^{1/2} = 1$ we find that ϕ satisfies

the given IVP. The solution is valid for $1 + \frac{2}{3}\ln(1+x^3) > 0$

or $x > (e^{-1.5} -1)^{1/3} \approx -0.91$.

5. Both solutions are valid for $c^2 > 4x^2$, which may be

    written $|x| < |c|/2$. The solution passing through $(1,-1)$

    is obtained by setting $x = 1$ in the second formula for y

    and solving for c, to yield $c = \sqrt{5}$.

6a. For $y_1 = 1-x$ then $y_1' = -1 = \dfrac{-x +[\ x^2 + 4(1-x)]^{1/2}}{2} =$

    $\dfrac{-x + [(x-2)^2]^{1/2}}{2} = \dfrac{-x + |x-2|}{2} = -1$ if $(x-2) \geq 0$ because of

    the definition of absolute value. Setting $x = 2$ in y_1 we

    get $y_1(2) = -1$ as required.

6b. By Theorem 2.2 we are guaranteed a unique solution

    only where $f(x,y) = \dfrac{-x +(x^2+4y)^{1/2}}{2}$ and $f_y(x,y) =$

    $(x^2+4y)^{-1/2}$ are continuous. In this case the initial

    point $(2,-1)$ lies in the region $x^2 + 4y \leq 0$, in which

    case $\dfrac{\partial f}{\partial y}$ is not continuous and hence the theorem is not

    applicable and there is no contradiction.

6c. If $y = y_2(x)$ then we must have $cx + c^2 = -x^2/4$, which is

    not possible since c must be a constant.

For problems 7 through 13 the isoclines are obtained by
setting the right side of the D.E. equal to a constant, and
then plotting the resulting curves for different values of c.

7. The isoclines are the circles

    $x^2+y^2 = c$, $c > 0$. This means that

    for $c = 1$ the solution will cross

    the circle $x^2+y^2 = 1$ with a

    slope of 1, while for $c = 2$ the

solution will cross the circle

$x^2+y^2 = 2$ with a slope of 2.

8. The isoclines are given by $x^2-xy + y^2-1 = c$. For $c>-1$,

    these are ellipses with center at the origin and with

    major axis on the line $y = x$.

9. Setting $xy/(1+x^2) = c$ and solving

    for y we get $y = c(x + 1/x)$

    as the isoclines.

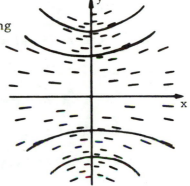

10. The isoclines are the straight lines $y = [(2-c)/(c+3)]x$.

12. The isoclines are given by

    $y(1-y^2) = c$, which are lines

    parallel to the x axis.

    For a given y value, say y_0 ,

    the solution curve will cross

    the isocline with a slope given

    by $c = y_0(1-y_0^2)$.

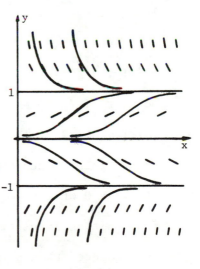

Section 2.4, Page 41

Problems 1 through 16 follow the pattern of the examples
worked in this section. The first eight problems, however, do

not have I.C. so the integration constant, c, cannot be found.

1.   Write the equation in the form y dy = x^2dx. Integrating

the left side with respect to y and the right side with

respect to x yields $\frac{y^2}{2} = \frac{x^3}{3} + c$. Since f(x,y) = x^2/y,

Theorem 2.2 is applicable in any region of the xy plane

for which y $\neq$ 0.

4.   Factor the right side to obtain y' = $(1+x)(1+y^2)$, which

separates to $(1+y^2)^{-1}$dy = (1+x)dx. Integrating each

side, as in problem 1, yields arctan y = $\frac{x^2}{2}$ + x + c, which

can be solved explicitly for y. A unique solution exists

everywhere since f(x,y) = $(1+x+y^2+xy^2)$ and its partial

derivative with respect to y are continuous everywhere.

6.   Separating the variables we get $(1-y^2)^{-1/2}$ dy = x^{-1}dx.

Integrating each side yields arcsin y = ln$|x|$+c, so y =

sin [ln$|x|$+c], x $\neq$ 0. The function f(x,y) =

$x^{-1}(1-y^2)^{1/2}$ is continuous for x $\neq$ 0 and $|y|{\leq}1$.

However, f_y(x,y) = $-yx^{-1}(1-y^2)^{-1/2}$ is continuous only for

$|y|$<1 and x $\neq$ 0. Thus we are guaranteed a unique solution

at all points such that x $\neq$ 0 and $|y|$<1. Since

f(x,y) = 0 when y = 1 or y = -1 and $\frac{dy}{dx}$ = 0,y = $\pm$ 1 are

two additional solutions. Theorem 2.2 is not

contradicted though, since these solutions lie outside

the region of applicability of the theorem.

9.   As above we start with cos3y dy = −sin 2x dx and

integrate to get $\frac{1}{3}$ sin 3y = $\frac{1}{2}$ cos 2x + c. Setting y =

π/3 when x = π/2 (from the I.C.) we find that 0 = $-\frac{1}{2}$ + c

or c = $\frac{1}{2}$, so that $\frac{1}{3}$ sin 3y = $\frac{1}{2}$ cos 2x + $\frac{1}{2}$ = $\cos^2 x$ (using

appropriate trigonometric identity). Solving for y

yields y = $\frac{1}{3}$ arcsin($3\cos^2 x$), which is defined only for 0 ≤

$3\cos^2 x \le 1$, or $-\sqrt{1/3} \le \cos x \le \sqrt{1/3}$. Taking the indicated

square roots and then finding the inverse cosine of each

side yields .955 ≤ x ≤ 2.19 as the approximate interval.

12. Separate variables by factoring the denominator of the

right side to get y dy = $\frac{2x}{1+x^2}$ dx. Integration yields

$y^2/2$ = $\ln(1+x^2)$+c and use of the I.C. gives c = 2. Thus

y = ± $[2\ln(1+x^2)+4]^{1/2}$, but we must discard the plus

square root because of the I.C. Since 1 + x^2 > 0, the

solution is valid for all x.

14. Separating variables and integrating yields y + y^2 =

x^2 + c. Setting y = 0 when x = 2 yields c = −4 or

y^2 + y = x^2−4. To solve for y complete the square on

the left side by adding 1/4 to both sides. This yields

y^2 + y + $\frac{1}{4}$ = x^2 −4 + $\frac{1}{4}$ or $(y + \frac{1}{2})^2$ = x^2−15/4. Taking

the square root of both sides yields y + $\frac{1}{2}$ = $\pm\sqrt{x^2 - 15/4}$,

where the positive square root must be taken in order to

satisfy the I.C. Thus y = $-\frac{1}{2}$ + $\sqrt{x^2 - 15/4}$, which is

defined for $x^2 \geq 15/4$ or $x \geq \sqrt{15/2}$. The possibility that
$x < -\sqrt{15/2}$ is discarded due to the I.C.

18. Assume $cx + d \neq 0$ and $c \neq 0$, then use long division to

obtain $\dfrac{dy}{dx} = \dfrac{a}{c} + (b - \dfrac{ad}{c}) \dfrac{1}{cx+d} = \dfrac{a}{c} + (\dfrac{bc-ad}{c}) \dfrac{1}{cx+d}$.

Integration then yields $y = \dfrac{a}{c} x + (\dfrac{bc-ad}{c^2}) \ln |cx + d| + k$.

20. If $v = y/x$ then $y = vx$ and $\dfrac{dy}{dx} = v + x\dfrac{dv}{dx}$ and thus the D.E.

becomes $v + x \dfrac{dv}{dx} = \dfrac{vx-4x}{x-vx} = \dfrac{v-4}{1-v}$. Subtracting v from both

sides yields $x\dfrac{dv}{dx} = \dfrac{v^2-4}{1-v}$, which separates into

$\dfrac{1-v}{v^2-4} dv = \dfrac{1}{x} dx$. To integrate the left side use

partial fractions to write $\dfrac{1-v}{v-4} = \dfrac{A}{v-2} + \dfrac{B}{v+2}$, which

yields $A = -1/4$ and $B = -3/4$. Integration then gives

$-\dfrac{1}{4} \ln|v-2| - \dfrac{3}{4} \ln|v+2| = \ln|x| - k$, or

$\ln|x^4||v-2||v+2|^3 = 4k$ after manipulations using

properties of the ln function. Setting $v = y/x$ and

further algebraic manipulations yield $(y-2x)(y+2x)^3 = c$,

where $c = e^{4k}$.

21a. The maximum and minimum values of $y/(1 + 2y^2)$ occur at

$y = +\sqrt{1/2}$ and $y = -\sqrt{1/2}$ respectively. Thus

$-\dfrac{1}{2\sqrt{2}} \leq \dfrac{y}{1+2y^2} \leq \dfrac{1}{2\sqrt{2}}$ and we know that $|\cos x| \leq 1$, hence

$f(x,y)$ is bounded as stated.

21b. Since $dy/dx = f(x,y)$, we have from part (a)$-\dfrac{1}{2\sqrt{2}} \leq \dfrac{dy}{dx} \leq \dfrac{1}{2\sqrt{2}}$

Separating variables and using the format of Eq.(13) of

this section yields $-\displaystyle\int_0^x \dfrac{dt}{2\sqrt{2}} \leq \int_1^\phi dy \leq \int_0^x \dfrac{dt}{2\sqrt{2}}$.

Integration and evaluation gives $\dfrac{-x}{2\sqrt{2}} \leq \phi(x) -1 \leq \dfrac{x}{2\sqrt{2}}$ or

$|\phi(x) -1| \leq \dfrac{|x|}{2\sqrt{2}}$ for all x. Geometrically, this says that

the solution is bounded by two straight lines of slope

$\pm\dfrac{1}{2\sqrt{2}}$ passing through the initial point x = 0, y = 1.

Thus the solution is bounded for all finite values of x.

Section 2.5, Page 50

1a. Comparing this problem with Eq.(1), we see that r = .0525

and thus Eq.(8) yields $\tau < \ln 2/.0525 = 13.20$ years as the

half life of plutonium 241.

1b. Solving $dQ/dt = -0.0525\,Q$ with $Q(0) = 50$ mg we find $Q(t)=$

$50e^{-0.0525t}$. Thus $Q(10) = 50e^{-0.525} = 29.6$ mg.

3. From Eq.(8), if $\tau = 1620$ years, then $r = \dfrac{\ln 2}{1620} =$

.000428 $(\text{years})^{-1}$. Thus, the solution of Eq.(1), with

$Q(0) = Q_0$ is $Q(t) = Q_0\, e^{-0.000428t}$. Setting $Q(t) = \dfrac{3}{4} Q_0$

and solving for t we obtain: $t = \dfrac{-\ln 3/4}{0.000428} = 672$ years.

4a. From Ex.(1) r = 0.02828 and thus $dQ/dt = -0.02828\,Q + 1$

gives the rate of change of thorium 234 in mg/day. The

solution of this equation with the initial condition of

$Q(0) = 100$ mg, can be found using methods of Section 2.1

and is $Q(t) = (100 - \frac{1}{0.02828})e^{-0.02828t} + \frac{1}{0.02828}$, or

$Q(t) = 64.64e^{-0.02828t} + 35.36$ mg.

4c. Setting $Q(t) = 13.86$ mg in the answer to part (a) and

solving for t yields $t = \frac{-\ln(.5/64.64)}{0.02828} = 171.9$ days.

4d. The D.E. for this problem is $dQ/dt = -0.02828Q + k$, which

has the solution $Q(t) = ce^{-0.02828t} + \frac{k}{0.02828}$ using

methods of Section 2.1. If $Q(0) = 100$, then $c + \frac{k}{0.02828} =$

100, or $c = 100 - \frac{k}{0.02828}$. Thus $Q(t) =$

$(100 - \frac{k}{0.02828})e^{-0.02828t} + \frac{k}{0.02828}$. If $Q(t)$ is to be at

a constant level of 100 mg, then

$100 - \frac{k}{0.02828} = 0$, or $k = 2.828$ mg/day.

6a. This problem is the same as the investment problem given

by Eqs.(13) and (14) with $k = 0$. Thus setting $k = 0$,

$t = T$ and $S(T) = 2S_0$ in Eq.(16) and solving for T we get

$T = \ln 2/r$ years.

6b. Set $r = .07$ in the answer to (a).

6c. Set $T = 8$ in the answer to (a) and then solve for r.

7a. Set $S_0 = 0$ in Eq.(16) or solve Eq.(13) with $S(0) = S_0$.

7b. Set $r = .075$, $t = 40$ and $S(t) = \$1,000,000$ in the answer

to (a) and then solve for k.

7c. Set $k = \$2,000$, $t = 40$ and $S(t) = \$1,000,000$ in the answer to (a) and then solve numerically for r.

10a. Since withdrawals are made, the rate of change of S(t) is given by $dS/dt = rS - k$ dollars/year. The solution to this equation is $S(t) = ce^{rt} + \frac{k}{r}$. If $S(0) = S_0$, then $c + \frac{k}{r} = S_0$, or $c = S_0 - \frac{k}{r}$ and thus $S(t) = (k/r) + [S_0-(k/r)]e^{rt}$.

10b. If S(t), as found in (a), is to be constant then the coefficient of e^{rt} must be zero. Thus $S_0 - k_0/r = 0$ or $k_0 = rS_0$.

10c. Setting $S(T) = 0$ in (a), we have $[S_0 - (k/r)]e^{rT} = - k/r$ or $T = \frac{1}{r} \ln \frac{k/r}{k/(r-S)} = \frac{1}{r} \ln \frac{k}{k-k_0}$ years.

10e. Set $t = T$ and $S(T) = 0$ in the answer to (a) and solve for k.

10f. Set $r = .08$, $k = \$12,000$, $t = 20$ and $S(20) = 0$ in the answer to (a) and solve for S_0.

12. Let p(t) be the population at any time and let k be the constant of proportionality. Then the rate of change of p(t) is given by $dp/dt = kp$, with $p(0) = 6 \times 10^8$ and $p(300) = 2.8 \times 10^9$. The solution of the D.E. is $p(t) = ce^{kt}$, where c is found by setting $t = 0$: $p(0) = c = 6 \times 10^8$. Thus $p(t) = (6 \times 10^8)e^{kt}$. To find k, set $t = 300$ and $p(300) = 2.8 \times 10^9$, yielding: $e^{300k} = (2.8/6) \times 10$ or

$k = .005135$. Hence $p(t) = (6 \times 10^8)e^{.005135t}$ gives the population at any time t. Setting $p(t) = 2.5 \times 10^{10}$ and solving for t yields the number of years after 1650 when the greatest population is reached.

13. From Eq.(22) we have $190^O = 70^O + (200^O-70^O)e^{-k}$ or $k =$
     $-\ln \dfrac{120}{130} = \ln 13/12$. Hence if $\theta(T) = 150^O$ we get: $150^O = 70^O + 130^O e^{-\ln(13/12)T}$. Solving for T yields $T = \ln(8/13)/-\ln(13/12) = \ln(13/8)/\ln(13/12)$.

15. To determine the cooling rate, the information relevant to the morgue is used in Eq.(22):
     $\theta(t) = 40^O + (85^O-40^O)e^{-kt}$. Setting $t = 1$ and $Q(1) = 60$ and solving for k yields $k = .8109(\text{hour})^{-1}$. To find the time of death, use Eq.(22) with $T = 70^O$ and $\theta_0 = 98.6^O$:
     $\theta(t) = 70^O + 28.6e^{-.8109t}$. Setting $t = T$ and $\theta(T) = 85$ and solving for T gives $T = .7959$ hours before midnight, or 11:12 P.M.

16. The D.E. expressing the evaporation is $dV/dt = kS$, where the volume $V = \dfrac{4}{3}\pi r^3$ and the surface area $S = 4\pi r^2$. The D.E., in terms of r, then is $r^2 \dfrac{dr}{dt} = kr^2$, with $r(0) = 3$ and $r(1) = 2$. When $r \neq 0$ the D.E. becomes $r' = k$, or $r(t) = kt + c$. Recall that r must be positive for the problem to have physical meaning.

18. In order to find the amount of salt at the end of 20

minutes, we must first find the amount of salt that is

present after 10 minutes. For the first 10 minutes (if

we let $Q(t)$ be the amount of salt in the tank):

$\frac{dQ}{dt} = \frac{1}{2}(2) - 2\frac{Q(t)}{100}$, $Q(0) = 0$. This I.V.P. has the

solution: $Q(t) = 50(1-e^{-.02t})$, which yields $Q(10) =$

9.063 lbs. of salt in the tank after the first 10

minutes. At this point no more salt is allowed to enter,

so the new I.V.P. (letting $P(t)$ be the amount of salt in

the tank) is: $\frac{dP}{dt} = (0)(2) -2\frac{P(t)}{100}$, $P(0) = Q(10) = 9.063$.

The solution of this problem is $P(t) = 9.063e^{-.02t}$, which

yields $P(10) = 7.42$ lbs. present 20 minutes after the

start.

19. Salt flows into the tank at the rate of (1)(3) lb/min.

and it flows out of the tank at the rate of

$\frac{Q(t)}{200+t}(2)$ lb/min. since the volume of water in the tank

at any time t is 200 + (1)(t) gallons due to the fact

that water flows into the tank faster than it flows out.

Thus the I.V.P. is $dQ/dt = 3 - \frac{2}{200+t} Q(t)$, $Q(0) = 100$.

20. Let $Q(t)$ be the quantity of carbon monoxide in the room.

Then $dQ/dt = (.04)(.1) - \frac{Q(t)}{1200} (.1)$, $Q(0) = 0$ is the

I.V.P. to be solved. Note that $x(t) = Q(t)/1200$.

21a. The required I.V.P. is $dQ/dt = kr + P - \frac{Q(t)}{V} r$, $Q(0) =$

Vc_0 . Since $c = Q(t)/V$, the I.V.P. may be rewritten

$Vc'(t) = kr + P - rc$, $c(0) = c_0$, which has the solution

$c(t) = k + \dfrac{P}{r} + (c_0 - k - \dfrac{P}{r})e^{-rt/V}$.

21b. Set $k = 0$, $P = 0$, $t = T$ and $c(T) = .5c_0$ in the solution

found in (a).

Section 2.6, Page 65

Problems 1 through 13 follow the pattern illustrated in Fig.2
and the discussion following Eq.(9).

3.   The critical points are found by setting $\dfrac{dN}{dt}$ equal to

zero. Thus $N = 0,1,2$ are the critical points. The graph

of $N(N-1)(N-2)$ is positive for $0 < N < 1$ and $2 < N$ and

negative for $1 < N < 2$. Thus $N(t)$ is increasing $(\dfrac{dN}{dt} > 0)$

for $0 < N < 1$ and $2 < N$ and decreasing $(\dfrac{dN}{dt} < 0)$ for $1 < N < 2$.

Therefore 0 and 2 are unstable critical points while 1 is

a stable critical point.

6. $\dfrac{dN}{dt}$ is zero only when Arctan N is zero. $\dfrac{dN}{dt} > 0$ for $N < 0$

and $\dfrac{dN}{dt} < 0$ for $N > 0$. Thus $N = 0$ is a stable critical

point.

7c. Separate variables to get $\dfrac{dN}{(1-N^2)} = kt$. Integration

yields $\dfrac{1}{1-N} = kt+c$, or $N = 1 - \dfrac{1}{kt+c} = \dfrac{kt+c-1}{kt+c}$.

Setting $t = 0$ and $N(0) = N_0$ yields $N_0 = \dfrac{c-1}{c}$ or $c = \dfrac{1}{1-N_0}$,

Hence $N(t) = \dfrac{(1-N_0)\, kt + N_0}{(1-N_0)\, kt + 1}$.

9.   Setting $\dfrac{dN}{dt} = 0$ we find $N = 0, \pm 1$ are the critical

points. Since $\frac{dN}{dt} > 0$ for $N < -1$ and $N > 1$ while $\frac{dN}{dt} < 0$ for

$-1 < N < 1$ we may conclude that $N = -1$ is stable, $N = 0$ is

semistable, and $N = 1$ is unstable.

11. $N = b^2/a^2$ and $N = 0$ are the only critical points. For

$0 < N < b^2/a^2$, $\frac{dN}{dt} < 0$ and thus $N = 0$ is a stable equilibrium

point. For $N > b^2/a^2$, $dN/dt > 0$ and thus $N = b^2/a^2$ is

unstable.

14. If $F'(N_1) < 0$ then the slope of N_1 is negative at N_1 and

thus $F(N) > 0$ for $N < N_1$ and $F(N) < 0$ for $N > N_1$ since

$F(N_1) = 0$. Hence N_1 is a stable critical point. A

similar argument holds for $F'(N_1) > 0$.

15. This problem is just like Eq.(6) and thus you may use

Eq.(9) with the appropriate values substituted in.

16b. The graph of $\frac{dN}{dt}$ vs N has a maximum point at $N = K/e$.

Thus $\frac{dN}{dt}$ is positive and increasing for $0 < N < K/e$ and thus

$N(t)$ is concave up for that interval. Similarly $\frac{dN}{dt}$ is

positive and decreasing for $K/e < N < K$ and thus $N(t)$ is

concave down for that interval.

16c. $\ln(K/N)$ is very large for small values of N and thus

$rN \ln (K/N) > r N(1-N/K)$ for small N. Since $\ln(K/N)$ and

$(1-N/K)$ are both strictly decreasing functions of N and

since $\ln(K/N) = (1-N/K)$ only for $N = K$, we may conclude that $\frac{dN}{dt} = rN \ln(K/N)$ is never less than $\frac{dN}{dt} = r N(1-N/K)$.

17a. If $u = \ln(N/K)$ then $N = Ke^u$ and $\frac{dN}{dt} = Ke^u \frac{du}{dt}$ so that the D.E. becomes $du/dt = -ru$.

18b. Use the results of Problem 14.

18d. Differentiate Y with respect to E.

19a. Set $\frac{dN}{dt} = 0$ and solve for N using the quadratic formula.

19b. Use the results of Problem 14.

19d. If $h > rk/4$ there are no critical points (see part a) and $\frac{dN}{dt} < 0$ for all t.

19e. See part (a).

22a. If $z = x/n$ then $dz/dt = \frac{1}{n} \frac{dx}{dt} - \frac{x}{n^2} \frac{dn}{dt}$. Use of Equations (i) and (ii) then give the I.V.P. (iii).

22b. Separate variables to get $\frac{dz}{z(1-z)} = -\beta dt$. Using partial fractions this becomes $\frac{dz}{z} + \frac{dz}{1-z} = -\beta dt$. Integration and solving for z yields the answer.

24a. Plot dx/dt vs x and observe that $x = p$ and $x = q$ are critical points. Also note that $dx/dt > 0$ for $x < \min(p,q)$ and $x > \max(p,q)$ while $dx/dt < 0$ for x between $\min(p,q)$ and $\max (p,q)$. Thus $x = \min(p,q)$ is a stable point while $x = \max(p,q)$ is unstable. To solve the D.E., separate variables and use partial fractions to obtain

$$\frac{1}{q-p} \left[\frac{dx}{q-x} - \frac{dx}{p-x} \right] = \alpha \, dt.$$

Integration and solving for x yields the solution.

24b. $x = p$ is a semistable critical point and since $\frac{dx}{dt} > 0$,

x(t) is an increasing function. Thus for $x(0) = 0$, $x(t)$

approaches p as $t \to \infty$. To solve the D.E., separate

variables and integrate.

Section 2.7, Page 76

1a. If $x(t)$ is the height above the ground, then the I.V.P.

for $v(t)$ is $dv/dt = -g$, $v(0) = 20$. Thus $\frac{dx}{dt} = v(t) = 20 -$

gt and $x(t) = 20t-(g/2)t^2 + c$. Since $x(0) = 30$, $c = 30$

and $x(t) = 20t - (g/2)t^2+30$. At the maximum height

$v(t_m) = 0$ and thus $t_m = 20/9.8 = 2.04$ sec., which when

substituted in the equation for $x(t)$ yields the maximum

height.

1b. At the ground $x(t_g) = 0$ and thus $20t_g -4.9t_g^2+30 = 0$.

2. The I.V.P. in this case is $m\frac{dv}{dt} = -\frac{1}{30}v - mg$, $v(0) = 20$,

where the positive direction is measured upward.

4. If the positive direction is taken as down, then the

I.V.P. for this problem is $m(dv/dt) = mg-kv$, $v(0) = v_0$.

5. The I.V.P. is $m(dv/dt) = mg-kv$, $v(0) = 0$ which has the

solution $v(t) = \frac{mg}{k}(1-e^{-kt/m})$. Setting $v(t) = .9\frac{mg}{k}$ and

solving for t yields the answer.

6. The I.V.P. is $m(dv/dt) = 10-2v$, $v(0) = 0$ with the

positive direction being the direction of motion of the

boat.

7a. The I.V.P. is $m \frac{dv}{dt} = mg - .75v$, $v(0) = 0$ and v is

measured positively downward. The solution to this

equation is $v(t) = 240(1-e^{-.133t})$ so that $v(10) = 176.7$

ft/sec.

7b. Integration of $v(t)$ as found in (a) yields $x(t) = 240t +$

$1800(e^{-.133t}-1)$ where x is measured positively down from

the altitude of 5000 feet. Set $t = 10$ to find the

distance traveled when the parachute opens.

7c. After the parachute opens the I.V.P. is $m \frac{dv}{dt} = mg-12v$,

$v(0) = 176.7$, which has the solution $v(t) =$

$161.7e^{-2.133t} + 15$ and where $t = 0$ now represents the

time the parachute opens. Letting $t \rightarrow \infty$ yields the

limiting velocity.

7d. Integrate $v(t)$ as found in (c) to find $x(t) =$

$15t - 75.8e^{-2.133t} + C_2$. $C_2 = 75.8$ since $x(0) = 0$, x now

being measured from the point where the parachute opens.

Setting $x = 3925.5$ will then yield the length of time the

skydiver is in the air after the parachute opens.

8.   The I.V.P. is $m(dv/dt) = mg - k\sqrt{v}$; $v(0) = v_0$ where the

positive direction is down. To solve, separate variables

to obtain $\frac{mdv}{mg-k\sqrt{v}} = dt$ and let $u = mg - k\sqrt{v}$ to integrate the left side.

9. The I.V.P. is $m(dv/dt) = mg - kv^2$, $v(0) = 0$. To solve, separate variables and use partial fractions.

12b. From part (a) $v(t) = -\frac{mg}{k} + [v_0 + \frac{mg}{k}]e^{-kt/m}$. As $k \to 0$ this has the indeterminant form of $-\infty + \infty$. Thus rewrite $v(t)$ as $v(t) = [-mg + (v_0 k + mg)e^{-kt/m}]/k$ which has the indeterminant form of $0/0$, as $k \to 0$ and hence L'Hopital's Rule may be applied with k as the variable.

13a. The equation of motion is $m(dv/dt) = w-R-B$ which, in this problem, is $\frac{4}{3}\pi a^3 \rho(dv/dt) = \frac{4}{3}\pi a^3 \rho g - 6\pi \mu a v - \frac{4}{3}\pi a^3 \rho' g$. The limiting velocity occurs when $dv/dt = 0$.

13b. Since the droplet is motionless, $v = dv/dt = 0$, we have the equation of motion $0 = (\frac{4}{3})\pi a^3 \rho g - Ee - (\frac{4}{3})\pi a^3 \rho' g$, where ρ is the density of the oil and ρ' is the density of air. Solving for e yields the answer.

14. At maximum altitude $v = 0$ and thus Eq. (16) may be solved for x when $v = 0$. Recall that $v_e = (2gR)^{1/2}$.

15b. Note that 32 ft/sec^2 = $78,545$ m/hr^2.

16. This problem is the same as Example 2 through Eq.(15). In this case the I.C. is $v(\xi R) = v_0$, so $c = v_o^2 - \frac{2gR}{1+\xi}$. v_e is found by noting that $v_o^2 > \frac{2gR}{1+\xi}$ in order for v^2 to always be positive. From Example 2, the escape

velocity for a surface launch is $V_e(0) = 2gR$. We want

the escape velocity at $x_o = R$ to have the relation

$v_e(\xi R) = .85 \, v_e(0)$, which yields $\xi = (0.85)^{-2} - 1 \simeq$

0.384. If $R = 4000$ miles then $x_o = \xi R = 1536$ miles.

Section 2.8, Page 83

3. $M(x,y) = 3x^2 - 2xy + 2$ and $N(x,y) = 6y^2 - x^2 + 3$, so $M_y =$

$-2x = N_x$ and thus the D.E. is exact. Integrating $M(x,y)$

with respect to x we get $\psi(x,y) = x^3 - x^2 + 2x + h(y)$. Taking

the partial derivative of this with respect to y

and setting it equal to $N(x,y)$ yields $-x^2 + h'(y) =$

$6y^2 - x^2 + 3$, so that $h'(y) = 6y^2 + 3$ and $h(y) = 2y^3 + 3y$.

Substitute this $h(y)$ into $\psi(x,y)$ and recall that the

equation which defines $y(x)$ implicitly is $\psi(x,y) = c$.

Thus $x^3 - x^2 y + 2x + 2y^3 + 3y = c$ is the equation that

yields the solution.

5. Writing the equation in the form $M(x,y)dx + N(x,y)dy = 0$

gives $M(x,y) = ax + by$ and $N(x,y) = bx + cy$. Now $M_y = b =$

N_x and the equation is exact. Integrating $M(x,y)$ with

respect to x yields $\psi(x,y) = (a/2)x^2 + bxy + h(y)$.

Differentiating ψ with respect to y (x constant) and

setting $\psi_y(x,y) = N(x,y)$ we find that $h'(y) = cy$ and

thus $h(y) = (c/2)y^2$. Hence the solution is given by

$(a/2)x^2 + bxy + (c/2)y^2 = k.$

6.   The D.E. must be put into the form of Eq.(3) or Eq.(7).

7. $M_y(x,y) = e^x\cos y - 2\sin x = N_x((x,y)$ and thus the D.E. is exact. Integrating $M(x,y)$ with respect to x gives $\psi(x,y) = e^x\sin y + 2y\cos x + h(y)$. Finding $\psi_y(x,y)$ from this and setting that equal to $N(x,y)$ yields $h'(y) = 0$ and thus $h(y)$ is a constant. Hence an implicit solution of the D.E. is $e^x\sin y + 2y \cos x = c$. The solution $y = 0$ is also valid since it satisfies the D.E. for all x.

9.   If you try to find $\psi(x,y)$ by trying to integrate $M(x,y)$ with respect to x you must integrate by parts. Instead find $\psi(x,y)$ by integrating $N(x,y)$ with respect to y to obtain $\psi(x,y) = e^{xy}\cos 2x - 3y + g(x)$. Now find $g(x)$ by differentiating $\psi(x,y)$ with respect to x and set that equal to $M(x,y)$, which yields $g'(x) = 2x$ or $g(x) = x^2$.

12.   As long as $x^2 + y^2 \neq 0$, we can simplify the equation by multiplying both sides by $(x^2 + y^2)^{3/2}$. This gives the exact equation $xdx + ydy = 0$. The solution to this equation is given implicitly by $x^2 + y^2 = c$. If you apply Theorem 2.3 and its construction without the simplification, you get $(x^2 + y^2)^{-1/2} = C$ which can be written as $x^2 + y^2 = c$ under the same assumption required for the simplification.

14. $M_y = 1$ and $N_x = 1$, so D.E. is exact. Integrating $M(x,y)$

with respect to x yields $\psi(x,y) = 3x^3 + xy - x + h(y)$.

Differentiating this with respect to y and setting

$\psi_y(x,y) = N(x,y)$ yields $h'(y) = -4y$ or $h(y) = -2y^2$.

Thus the implicit solution is $3x^3 + xy - x - 2y^2 = c$.

Setting $x = 1$ and $y = 0$ gives $c = 2$ so that

$2y^2 - xy + (2+x-3x^3) = 0$ is the implicit solution

satisfying the given I.C. Use the quadratic formula to

find y(x), where the negative square root is used in

order to satisfy the I.C.

15a. We want $M_y(x,y) = 2xy + bx^2$ to be equal to $N_x(x,y) =$

$3x^2 + 2xy$. Thus we must have $b = 3$. This gives

$\psi(x,y) = \frac{1}{2} x^2 y^2 + x^3 y + h(y)$ and $h'(y) = 0$ so, after

multiplying through by 2, the solution is given

implicitly by $x^2 y^2 + 2x^3 y = c$.

15b. Using similar procedures, we must solve $e^{2xy} + 2xye^{2xy} =$

$be^{2xy} + 2bxye^{2xy}$ for b, which gives $b = 1$. Solving the

equation with $b = 1$, we find that $e^{2xy} + x^2 = c$.

Section 2.9, Page 87

1. $M_y(x,y) = 3x^2 y^2$ and $N_x(x,y) = 1 + y^2$ so the equation is

not exact by Theorem 2.3. Proceeding as in Example 1

and multiplying by the integrating factor $\mu(x,y) = 1/xy^3$

we get $x + \frac{(1+y^2)}{y^3} y' = 0$ which is an exact equation since

$M_y = N_x = 0$ (it is also separable). In this case $\psi =$
$\frac{1}{2}x^2 + h(y)$ and $h'(y) = y^{-3} + y^{-1}$ so that
$x^2 - y^{-2} + 2\ln|y| = c$ gives the solution implicitly.

4. Multiplication of the given D.E. (which is not exact) by

$\mu(x,y) = xe^x$ yields $(x^2 + 2x)e^x$ sin y dx $+ x^2e^x$ cos y dy,

which is exact since $M_y(x,y) = N_x(x,y) = (x^2+2x)e^x$ cos y.

To solve this exact equation it's easiest to integrate

$N(x,y) = x^2e^x$ cos y with respect to y to yield $\psi(x,y) =$

x^2e^x sin y $+ g(x)$. Solving for $g(x)$ yields the implicit

solution.

5. This problem is similar to the derivation leading up to

Eq. (7). Assuming that μ depends only on y, we find from

Eq. (4) that $\mu' = Q\mu$, where $Q = (N_x - M_y)/M$ must depend

on y alone. In this case the integrating factor is as

given. This provides an alternative approach to Problems

9 through 12.

7. The equation is not exact so we must attempt to find an

integrating factor. Since $\frac{1}{N}(M_y - N_x) = \dfrac{3x^2 + 2x + 3y^2 - 2x}{x^2 + y^2}$

$= 3$ is a function of x alone there is an integrating

factor depending only on x, as shown in Eq. (7). Then

$d\mu/dx = 3\mu$, and the integrating factor is $\mu(x) = e^{3x}$.

Hence The equation can be solved as in Example 2.

8.   An integrating factor can be found using the method of

     Eq. (7). We find that $\frac{d\mu}{dx} = -\mu$ and hence $\mu(x) = e^{-x}$.

     Alternatively, you might recognize that $y' - y = e^{2x} - 1$

     is a linear first order equation which can be solved as

     in Sections 2.1 and 2.2 of the text.

9.   Using the results of Problem 5, it can be shown that

     $\mu(y) = y$ is an integrating factor. Thus multiplying the

     D.E. by y gives $y\,dx + (x - y \sin y)dy = 0$, which can be

     identified as an exact equation. Alternatively, one can

     rewrite the last equation as $(y\,dx + x\,dy) - y \sin y\,dy = 0$.

     0.   The first term is $d(xy)$ and the last can be

     integrated by parts. Thus we have $xy + y \cos y - \sin y = c$.

11. By multiplying by sin y we obtain $e^{x}\sin y\,dx + e^{x}\cos y\,dy + 2y\,dy = 0$, and the first two terms are just

     $d(e^{x}\sin y)$. Thus $e^{x}\sin y + y^2 = c$.

13. Using the results of Problem 6, it can be shown that

     $\mu(xy) = xy$ is an integrating factor. Thus multiplying by

     xy we have $(3x^2 y + 6x)dx + (x^3 + 3y^2)dy = 0$, which can be

     identified as an exact equation. Alternatively, we can

     observe that the above equation can be written as

     $d(x^3 y) + d(3x^2) + d(y^3) = 0$, so that $x^3 y + 3x^2 + y^3 = c$.

Section 2.10, Page 90

Problems 1 through 8 are done in the same manner as the
illustrative example of this section.

1.   The D.E. can be written as $\frac{dy}{dx} = 1 + y/x$ and is thus

     homogeneous. Setting $v = y/x$ yields $x\frac{dv}{dx} + v = 1 + v$ so

     that $\frac{dv}{dx} = \frac{1}{x}$. This equation is separable and has the

     solution $\ln|x| = v + c$. Since $v = y/x$, we obtain

     $x\ln|x| = y + cx$ as the solution.

3.   Writing the equation so that the right side is a

     function of y/x gives $dy/dx = 1 + (y/x) + (y/x)^2$ so the

     equation is homogeneous. The substitution $y = vx$ leads

     to $v + x\frac{dv}{dx} = 1 + v + v^2$ or $\frac{dv}{1 + v^2} = \frac{dx}{x}$. Solving, we get

     $\arctan v = \ln|x| + c$. Substituting for v we obtain

     $\arctan (y/x) - \ln|x| = c$.

5.   Dividing the numerator and denominator of the right

     side by x and substituting $y = vx$ we get $v + x\frac{dv}{dx} = \frac{4v - 3}{2 - v}$

     which can be rewritten as $x\frac{dv}{dx} = \frac{v^2 + 2v - 3}{2 - v}$. Separating

     variables gives $\frac{2 - v}{(v+3)(v-1)} dv = \frac{1}{x}dx$. Applying a partial

     fraction decomposition to the left side we obtain

     $[\frac{1}{4}\frac{1}{v-1} - \frac{5}{4}\frac{1}{v+3}]dv = \frac{dx}{x}$, and upon integrating both

     sides we find that $\frac{1}{4}\ln|v-1| - \frac{5}{4}\ln|v+3| = \ln|x| + c$.

     Substituting for v and performing some algebraic

     manipulations we get the solution in the implicit form

     $|y-x| = c|y+3x|^5$.

8. First write the equation in the form $dy/dx = 1 + 3(y/x) +$
    $(y/x)^2$. Substituting $y = vx$ and separating variables, we
    obtain $\dfrac{dv}{(1+v)^2} = dx$ which leads to $-(1+v)^{-1} - \ln|x| = c$.
    Substituting for v gives $\dfrac{x}{x + y} + \ln|x| = c$.

9b. Making the suggested substitution we obtain $\dfrac{dY}{dx} =$
    $\dfrac{2(Y-k) - (X-h) + 5}{2(X-h) - (Y-k) - 4} = \dfrac{2Y-X+(h-2k+5)}{2X-Y+(k-2h-4)}$, which will be
    homogeneous in X,Y provided the two expressions in
    parenthesis are both zero. Solving these yields $h = -1$
    and $k = 2$. From part (a), it then follows that $|Y-X| =$
    $c|Y + X|^3$. Substituting $Y = y + 2$, $X = x - 1$ gives the
    solution in the form $|y - x + 3| = c|y + x + 1|^3$.

10. The transformation $y = Y - 1$, $x = X - 3$ reduces the
    equation to Problem 6.

12. As in previous problems we can write the D.E. as $x\dfrac{dv}{dx} =$
    $-\dfrac{2v(v+2)}{v+1}$. Separating variables, using a partial
    fraction decomposition, and solving gives $|v|^{1/2}|v+2|^{1/2}x^2$
    $= c$. Substituting for v and simplifying yields $x^2y^2 +$
    $2x^3y = c$ which is the same as the answer found in Example
    2, Section 2.9.

13. To show that $\mu(x,y)$ is an integrating factor, multiply
    both terms of the D.E. by the given $\mu(x,y)$. Since the
    D.E. is homogeneous, we know that $M(x,y)/N(x,y) = F(y/x)$

and thus the D.E. may be written as $\dfrac{F}{xF+y}$ dx $+$ $\dfrac{1}{xF+y}$ dy $= 0$

after multiplying both numerators and denominators by 1/N

and substituting F for M/N. It may now be shown that

$\dfrac{\partial}{\partial x}[\dfrac{1}{xF+y}]$ $=$ $\dfrac{\partial}{\partial y}[\dfrac{F}{xF+y}]$ and hence the equation is exact.

14b. Find $\mu(x,y) =$ $\dfrac{1}{x^3 + 3xy^2 - 2xy^2}$ $=$ $\dfrac{1}{x(x^2+y^2)}$ and

multiply the equation by the integrating factor to get

$\dfrac{x^2 + 3y^2}{x(x^2+y^2)}$ dx $-$ $\dfrac{2xy}{x(x^2+y^2)}$ dy $= 0$. This equation is exact

by the previous problem so $\psi_y(x,y) = -$ $\dfrac{2y}{x^2+y^2}$.

Integrating with respect to y gives $\psi(x,y) = -\ln(x^2+y^2)+$

$g(x)$. We must have $\psi_x(x,y) = -$ $\dfrac{2x}{x^2+y^2}$ $+ g'(x) =$

$\dfrac{x^2 + 3y^2}{x(x^2+y^2)}$. Solving for $g'(x)$ gives $g'(x) =$

$\dfrac{x^2 + 3y^2 + 2x^2}{x(x^2+y^2)} = \dfrac{3}{x}$, so $g(x) = 3 \ln x + \ln \hat{c}$. We finally

obtain $\psi(x,y) = -\ln(x^2 + y^2) + 3 \ln|x| + \ln \hat{c}$ or $x^3 =$

$c(x^2 + y^2)$.

Section 2.11, Page 92

In Problems 1 through 32 the equation is identified as to type.

1. Linear

2. Homogeneous

3. Exact

4. Linear equation in x

5. Exact

6. Linear

7. Linear equation in u   8. Linear

9. Exact   10. Integrating factor
                                      depends on x only

11. Exact   12. Linear

13. Homogeneous   14. Exact or homogeneous

15. Separable   16. Homogeneous

17. Linear   18. Linear or homogeneous

19. Integrating factor   20. Separable
     depends on x only

21. Homogeneous   22. Separable

23. Bernoulli equation   24. Separable

25. Exact   26. Integrating factor
                                      depends on x only

27. Integrating factor   28. Exact
     depends on x only

29. Homogeneous   30. Linear equation in x

31. Separable   32. Integrating factor
                                      depends on y only

34. Differentiating the equation with respect to x gives

$p = \dfrac{dy}{dx} = \dfrac{1}{p}\dfrac{dp}{dx}$. Separating variables gives $dx = \dfrac{1}{p^2}\,dp$

which has the solution $x = -\dfrac{1}{p} + c$. Hence the solution

may be represented parametrically by $y = \ln p$, $x =$

$-\dfrac{1}{p} + c$. To solve the equation in another manner write it

as $p = \dfrac{dy}{dx} = e^y$ which separates to $e^{-y}dy = dx$ and we get

the solution $x + e^{-y} = c$.

36b. Comparing the given D.E. with the D.E. of Problem 35 we

see that in this case $q_2(x) = -1/x$ and $q_3(x) = 1$. Hence,

using the method suggested in Problem 35 we must solve

$\frac{dv}{dx} = -(-1/x + 2y_1)v - 1$, where $y_1(x)$ is given as $1/x$.

This is a linear first-order equation which has the

solution $v(x) = (\frac{c-x^2}{2})/x$ and thus a general solution to

the Riccati equation is $y_2(x) = 1/x + 2x/(c-x^2)$.

39b. Following the procedure outlined in Problem 38, we first

find that the given family of curves has the slope

$dy/dx = \frac{-x+c}{y}$ at each point. To eliminate c, solve the

given D.E. for c to get $c = \frac{x^2 + y^2}{2x}$. Thus $dy/dx = \frac{y^2 - x^2}{2xy}$

and the orthogonal family of curves will then have slopes

given by $dy/dx = \frac{-2xy}{y^2-x^2}$. This equation may be written as

$y' = \frac{-2(y/x)}{(y/x)^2-1}$ so it is homogeneous. Substitution of $y =$

vx, separation of variables, and a partial fraction

decomposition lead to the equation $(\frac{1}{v} - \frac{2v}{1+v^2})dv = \frac{dx}{x}$

whose solution is given implicitly by $\frac{v}{1 + v^2} = \hat{c}x$.

Substitution for v and algebraic simplification give

$\frac{y}{x^2+y^2} = \hat{c}$ or $x^2 + y^2 = 2cy$. Completion of the square

gives the answer in more easily recognizable form.

39d. The solution $y + \sqrt{x^2+y^2} = \tilde{c} x^2$ is obtained if the

procedure used in Problem 39b is followed. This answer

can be put in the desired (and preferable) form by

subtracting y from both sides and then squaring to obtain

$x^2+y^2 = (\tilde{c} x^2-y)^2 = \tilde{c}^2 x^4 - 2\tilde{c} x^2y + y^2$.   Subtracting y^2

and division by x^2 ($x \neq 0$) yields the answer.

40b. The slope of the given family of curves is given by m_2 =

dy/dx = −x/y and tan θ = 1. Thus, from the given

relationship of θ , m_1 and m_2 we have $1 = \dfrac{m_1 + x/y}{1-m_1 (x/y)}$.

Solving for m_1, we find the intersecting family of curves

must have slopes m_1 = dy/dx = (y−x)/(y+x), which is a

homogeneous D.E.

41. Using the equation for the slope of a line passing

through the points (x,y) and (a,b) we find the equation

to solve is y' = (y−b)/(x−a).

43. In this case the point through which the tangent line

must pass is (0, y/2) and the equation to solve is y' =

y/2x.

44. The D.E. is $S' = kS^2$, subject to the conditions S(0) = 1

million, S(1) = 2 million. Separating variables and

integrating leads to $S(t) = - \dfrac{1}{kt + c}$. The I.C. S(0) = 1

gives c = −1. Using the second condition, S(1) = 2 =

$- \dfrac{1}{k - 1}$ gives $k = \dfrac{1}{2}$ so the solution is $S(t) = \dfrac{2}{2 - t}$

million. In 6 months, $t = \dfrac{3}{2}$ (since t = 0 occurred 1 year ago) and $S(\frac{3}{2})$ = 4 million. As t approaches 2, his wealth

approaches infinity, so the solution cannot be extended

until two years from now (which would be t = 3).

45a. The D.E. is dV/dt = k $- \alpha \pi r^2$. The volume of a cone of

height ℓ and radius r is given by V = $\pi r^2 \ell/3$ where ℓ =

hr/a from symmetry. Solving for r yields the desired

solution.

45b. From the material on logistic growth in Section 2.6 we

have equilibrium given by k $- \alpha \pi r^2$ = 0.

45c. $\ell \leq$ h.

CHAPTER 3

Section 3.1, Page 111

1a. Let $v = y'$, then $v' = y''$ and thus the D.E. becomes $x^2 v' + 2xv - 1 = 0$ or $x^2 v' + 2xv = 1$. The left side is recognized as $(x^2 v)'$ and thus we may integrate to obtain $x^2 v = x + c$ (otherwise, divide both sides of the D.E. by x^2 and find the integrating factor, which is just x^2 in this case). Solving for $v = dy/dx$ we find $dy/dx = 1/x + c/x^2$ so that $y(x) = \ln x + c_1/x + c_2$.

1c. Set $v = y'$, then $v' = y''$ and thus the D.E. becomes $v' + xv^2 = 0$. This equation is separable and has the solution $-v^{-1} + x^2/2 = c$ or $v = y' = -2/(c_1 - x^2)$ where $c_1 = 2c$. We must consider separately the cases $c_1 = 0$, $c_1 > 0$ and $c_1 < 0$. If $c_1 = 0$, then $y' = 2/x^2$ or $y(x) = -2/x + c_2$. If $c_1 > 0$, let $c_1 = k^2$. Then $y' = -2/(k^2 - x^2) = -(1/k)[1/(k-x) + 1/(k+x)]$, so that $y(x) = (1/k)\ln|(k-x)/(k+x)| + c_2$. If $c_1 < 0$, let $c_1 = -k^2$. Then $y' = 2/(k^2 + x^2)$ so that $y(x) = (2/k)\tan^{-1}(x/k) + c_2$. Finally, we note that $y = $ constant is also a solution of the D.E.

2a. Following the procedure outlined, let $v = dy/dx$ and $y'' = dv/dx = v\, dv/dy$. Thus the D.E. becomes $y\, v\, dv/dy + v^2 = 0$, which is a separable equation with the solution $v = $

c_1/y. Next let $v = dy/dx = c/y$, which again separates

to give the solution $y^2 = c_1 x + c_2$.

2d. Again let $v = y'$ and $v' = v \, dv/dy$ to obtain $2y^2 v \, dv/dy +$

$2yv^2 = 1$. This is an exact equation with solution $v =$

$\pm \, y^{-1}(y + c_1)^{1/2}$. To solve this equation, we write it in

the form $\pm \, y \, dy/(y+c_1)^{1/2} = dx$. On observing that the

left side of the equation can be written as

$\pm \, [(y+c_1) - c_1]dy/(y+c_1)^{1/2}$ we integrate and find $\pm \, (2/3) \cdot$

$(y-2c_1)(y+c_1)^{1/2} = x + c_2$.

2f. If $v = y'$, then $v' = v \, dv/dy$ and the D.E. becomes $v \, dv/dy$

$+ \, v^2 = 2e^{-y}$. Dividing by v we obtain $dv/dy + v =$

$2v^{-1}e^{-y}$, which is a Bernoulli equation. Let $w(y) = v^2$,

then $dw/dy = 2v \, dv/dy$ and the last D.E. becomes $dw/dy +$

$2w = 4e^{-y}$, which is linear in w. It's solution is $w = v^2$

$= c_1 e^{-2y} + 4e^{-y}$. Setting $v = dy/dx$, we obtain a

separable equation in y and x, which is solved to yield

the solution.

3a. Since both x and y are missing, either approach

illustrated in Problems 1 and 2 will work. In this case

using the approach of Problem 1 is easier, so let $v = y'$

and thus $v' = y''$ and the D.E. becomes $v \, dv/dx = 2$. This

equation is separable and has the solution $v = dy/dx =$

$\pm \, (4x + c)^{1/2}$. Since we know $y' = 2$ when $x = 0$, we may

solve for $c = 4$, noting that we must use the positive

square root only. Thus $dy/dx = 2(x + 1)^{1/2}$, which may be integrated to yield the solution.

3b. The variable x is missing, so use the approach of Problem 2.

3d. The variable y is missing. Let $v = y'$, then $v' = y''$. The D.E. becomes $vv' - x = 0$. The solution of the separable equation is $v^2 = x^2 + c_1$. Substituting $v = y'$ and applying the I.C. $y'(1) = 1$, we obtain $y' = x$. The positive square root was chosen because $y' > 0$ at $x = 1$. Solving this last equation and applying the I.C. $y(1) = 2$, we obtain $y = x^2/2 + 3/2$.

4c. Divide the D.E. by $x(x-1)$ to obtain $y'' + [3/(x-1)]y' + [4/x(x-1)]y = 2/x(x-1)$, which is in the standard form of Theorem 3.2, with $p(x) = 3/(x-1)$, $q(x) = 4/x(x-1)$ and $g(x) = 2/x(x-1)$. Theorem 3.2 guarantees a unique solution satisfying the I.C. at x_0 in the open interval for which p,q, and g are continuous. Since p is continuous everywhere except at $x = 1$ and q and g are continuous everywhere except at $x = 0$ and $x = 1$, we conclude there will be a unique solution on any interval not containing either $x = 0$ or $x = 1$.

5. Since $y = \phi(x)$ is a solution of the I.V.P., we have $\phi''(x) + p(x)\phi'(x) + q(x)\phi(x) = 0$ and $\phi(0) = a_0$, $\phi'(0) =$

a_1. The differential equation is satisfied for all x in the open interval containing the origin; hence, $\phi''(0)$ + p(0) $\phi'(0)$ + q(0) $\phi(0)$ = 0, and in particular $\phi''(0)$ = - $[a_1 p(0) + a_0 q(0)]$. To find $\phi'''(x)$, we differentiate the D.E. to obtain $\phi'''(x)$ + p(x) $\phi''(x)$ + p'(x) $\phi'(x)$ + q(x) $\phi'(x)$ + q'(x) $\phi(x)$ = 0. Solving for $\phi'''(0)$ and making use of the above expression for $\phi''(0)$, we obtain $\phi'''(0) = [p^2(0) - p'(0) - q(0)]a_1 + [p(0)q(0) - q'(0)]a_0$. Since the coefficients in the D.E. are polynomials, and hence are infinitely differentiable, the process may be continued indefinitely.

6c. We have y = $c_1 x + c_2$ sin x, y' = $c_1 + c_2$ cos x, y" = $-c_2$sin x. From the last equation $c_2 = - y''/(\sin x)$. Substituting this in the first equation gives c_1 = (y+y")/x. Finally, substituting for c_1 and c_2 in the second equation we obtain (1-x cot x)y" - xy' + y = 0.

Section 3.2, Page 120

3. We show by direct substitution that e^x and e^{-x} are solutions of the D.E. Since the D.E. is linear and homogeneous it follows from Theorem 3.3 that $\phi(x) = c_1 e^x$ + $c_2 e^{-x}$ is a solution for any values of c_1 and c_2. In particular sinh x = $(e^x - e^{-x})/2$ and cosh x = $(e^x + e^{-x})/2$ are also solutions of the D.E.

6. Substitute y = 1 into the D.E. to show that it is a

solution. Similarly for $y = x^{1/2}$. If $y = c_1(1) + c_2 x^{1/2}$ is substituted in the D.E. you will get $-c_1 c_2/4x^{3/2}$, which is zero only if $c_1 = 0$ or $c_2 = 0$. Thus the linear combination of two solutions is not, in general, a solution. Theorem 3.3 is not contradicted however, since the D.E. is not linear.

7. $y = \phi(x)$ is a solution of the D.E. so $L[\phi](x) = g(x)$. Since L is a linear operator, $L[c\phi](x) = cL[\phi](x) = cg(x)$. But, since $g(x) \neq 0$, $cg(x) = g(x)$ if and only if $c = 1$. This is not a contradiction of Theorem 3.3 since the linear D.E. is not homogeneous.

8b. $y = \sin x$, $y' = \cos x$, and $y'' = -\sin x$, so $L[\sin x] = -a \sin x + b \cos x + c \sin x = (c-a)\sin x + b \cos x$.

9b. $y = e^{rx}$, $y' = re^{rx}$, $y'' = r^2 e^{rx}$. Hence $L[e^{rx}] = ax^2(r^2 e^{rx}) + bx(re^{rx}) + c(e^{rx}) = (ax^2 r^2 + bxr + c)e^{rx}$.

10c.
$$W(x,xe^x) = \begin{vmatrix} x & xe^x \\ 1 & e^x + xe^x \end{vmatrix} = xe^x + x^2 e^x - xe^x = x^2 e^x.$$

11b. We verify by direct substitution that $y_1 = e^{-x}$ and $y_2 = e^{2x}$ are solutions of the D.E. for all x. Next, $W(e^{-x},e^{2x}) = 3e^x \neq 0$, $-\infty < x < \infty$. Hence y_1 and y_2 are a fundamental set of solutions for all x.

13b. In Problem 3 we verified that $y_1(x) = \sinh x$ and $y_2(x) = \cosh x$ are solutions of the D.E. for all x. Next,

$W(\sinh x, \cosh x) = -1 \neq 0$, $-\infty < x < \infty$. Thus, y_1 and y_2

form a fundamental set of solutions of the D.E. on

$-\infty < x < \infty$, and any solution can be written in the form

$\phi(x) = c_1 \sinh x + c_2 \cosh x$. The. I.C. are $\phi(0) = 0 \rightarrow$

$c_1 \sinh 0 + c_2 \cosh 0 = 0 \rightarrow c_2 = 0$, and $\phi'(0) = 1 \rightarrow$

$c_1 \cosh (0) = 1 \rightarrow c_1 = 1$. The solution of the D.E.

satisfying the I.C. is $\phi(x) = \sinh x$.

14. From Eq.(15) we have $W(y_1,y_2) = c\exp[-\int^x p(t)dt]$, where

$p(t) = 2/t$ from the D.E. Thus $W(y_1,y_2) = c/x^2$. Since

$W(y_1,y_2)(1) = 2$ we find $c = 2$ and thus $W(y_1,y_2)(5) =$

$2/25$.

15. Let c be in a point in (α,β) at which both y_1 and y_2

vanish. Then $W(y_1,y_2)(c) = y_1(c)y_2'(c) - y_1'(c)y_2(c) =$

0. Hence, from the discussion following Theorem 3.4 the

functions y_1 and y_2 cannot form a fundamental set.

17. Suppose that y_1 and y_2 have a point of inflection at x_0

and either $p(x_0) \neq 0$ or $q(x_0) \neq 0$. Since $y_1''(x_0) = 0$ and

$y_2''(x_0) = 0$ it follows from the D.E. that $p(x_0)y_1'(x_0) +$

$q(x_0)y_1(x_0) = 0$ and $p(x_0)y_2'(x_0) + q(x_0)y_2(x_0) = 0$. If

$p(x_0) = 0$ and $q(x_0) \neq 0$ then $y_1(x_0) = y_2(x_0) = 0$, and

$W(y_1,y_2)(x_0) = 0$ so the solutions cannot form a

fundamental set. If $p(x_0) \neq 0$ and $q(x_0) = 0$ then $y_1'(x_0)$

$= y_2'(x_0) = 0$ and $W(y_1,y_2)(x_0) = 0$, so again the

solutions cannot form a fundamental set. If $p(x_0) \neq 0$

and $q(x_0) \neq 0$ then $y_1'(x_0) = q(x_0)y_1(x_0)/p(x_0)$ and

$y_2'(x_0) = q(x_0)y_2(x_0)/p(x_0)$ and thus

$$W(y_1,y_2)(x_0) = y_1(x_0)y_2'(x_0) - y_1'(x_0)y_2(x_0)$$
$$= y_1(x_0)[q(x_0)y_2(x_0)/p(x_0)] -$$
$$[q(x_0)y_1(x_0)/p(x_0)]y_2(x_0)$$
$$= 0$$

18c. First, we derive the necessary condition that the D.E.

$P(x)y'' + Q(x)y' + R(x)y = 0$ be exact. Suppose that

$P(x)y'' + Q(x)y' + R(x)y = [P(x)y']' + [f(x)y]'$. On

expanding the right side and equating coefficients, we

find $f'(x) = R(x)$ and $P'(x) + f(x) = Q(x)$. These two

conditions on f can be satisfied if $R(x) = Q'(x) - P''(x)$

which gives the necessary condition $P''(x) - Q'(x) + R(x)$

$= 0$. For the particular problem (c) we have $P(x) = x$,

$Q(x) = - \cos x$, and $R(x) = \sin x$ and the condition for

exactness is satisfied. Also $f(x) = Q(x) - P'(x) =$

$- \cos x - 1$, so the D.E. becomes $(xy')' - [(1 + \cos x)y]'$

$= 0$. Hence $xy' - (1 + \cos x)y = c_1$. This is a first

order linear D.E. and the integrating factor (after

dividing by x) is $\mu(x) = \exp[- \int^x t^{-1}(1 + \cos t)dt]$. The

general solution is $y = [\mu(x)]^{-1}[c_1 \int^x t^{-1} \mu(t)dt + c_2]$.

19a. We want to choose $\mu(x)$ and $f(x)$ so that $\mu(x)P(x)y'' +$

$\mu(x)Q(x)y' + \mu(x)R(x)y = [\mu(x)P(x)y']' + [f(x)y]'$. Expand

the right side and equate coefficients of y'', y' and y.

This gives $\mu'(x)P(x) + \mu(x)P'(x) + f(x) = \mu(x)Q(x)$ and $f'(x) = \mu(x)R(x)$. Differentiate the first equation and then eliminate $f'(x)$ to obtain the adjoint equation $P\mu'' + (2P' - Q)\mu' + (P'' - Q' + R)\mu = 0$. For the Bessel equation of order ν, $P(x) = x^2$, $Q(x) = x$, and $R(x) = x^2 - \nu^2$. Hence the adjoint equation is $x^2\mu'' + 3x\mu' + (1 + x^2 - \nu^2)\mu = 0$.

20. Write the adjoint D.E. given in Problem 19 as $\hat{P}\mu'' + \hat{Q}\mu' + \hat{R}\mu = 0$ where $\hat{P} = P$, $\hat{Q} = 2P' - Q$, and $\hat{R} = P'' - Q' + R$. The adjoint of this equation, namely the adjoint of the adjoint, is $\hat{P}y'' + (2\hat{P}' - \hat{Q})y' + (\hat{P}'' - \hat{Q}' + \hat{R})y = 0$. After substituting for $\hat{P}$, $\hat{Q}$, and $\hat{R}$ and simplifying, we obtain $Py'' + Qy' + Ry = 0$. This is the same as the original equation.

21. From Problem 19 the adjoint of $Py'' + Qy' + Ry = 0$ is $P\mu'' + (2P' - Q)\mu' + (P'' - Q' + R)\mu = 0$. The two equations are the same if $2P' - Q = Q$ and $P'' - Q' + R = R$. This will be true if $P' = Q$. Hence the original D.E. is self-adjoint if $P' = Q$. For Problem 19(a), $P(x) = x^2$ so $P'(x) = 2x$ and $Q(x) = x$. Hence the Bessel equation of order ν is not self-adjoint. In a similar manner we find that (b) and (c) are self adjoint.

Section 3.3, Page 126

3. The functions can be shown to be solutions of the D.E. by

substituting directly into the D.E. To show that y_1 and y_2 are linearly independent we must show that the Wronskian does not vanish. $W(e^{-2x}, e^{3x}) = e^{-2x}(e^{3x})' - e^{3x}(e^{-2x})' = 5e^x \neq 0$ for any x, and thus y_1 and y_2 are linearly independent.

6. Since $\cos 3x = 4 \cos^3 x - 3 \cos x$ we have $\cos 3x - (4 \cos^3 x - 3 \cos x) = 0$ for all x. From Eq.(1) we have $k_1 = 1$ and $k_2 = -1$ and thus $\cos 3x$ and $4 \cos^3 x - 3 \cos x$ are linearly dependent.

10. The D.E. is linear and homogeneous. Hence, if y_1 and y_2 are solutions, then $y_3 = y_1 + y_2$ and $y_4 = y_1 - y_2$ are solutions. $W(y_3, y_4) = y_3 y_4' - y_3' y_4 = (y_1 + y_2)(y_1' - y_2') - (y_1' + y_2')(y_1 - y_2) = -2(y_1 y_2' - y_1' y_2) = -2W(y_1, y_2)$, is not zero since y_1 and y_2 are linearly independent solutions. Hence y_3 and y_4 form a fundamental set of solutions.

12. Let $-1 < x_0$, $x_1 < 1$ and $x_0 \neq x_1$. If $y_1 = x$ and $y_2 = x^2$ are linearly dependent then $c_1 x_1 + c_2 x_1^2 = 0$ and $c_1 x_0 + c_2 x_0^2 = 0$ have a solution for c_1 and c_2 such that c_1 and c_2 are not both zero. But this system of equations has a non-zero solution only if $x_1 = 0$ or $x_0 = 0$ or $x_1 = x_0$. Hence, the only set c_1 and c_2 that satisfies the system for every choice of x_0 and x_1 in $-1 < x < 1$ is $c_1 = c_2 = 0$. Therefore x and x^2 are linearly independent on $-1 < x < 1$.

Next, $W(x,x^2) = x^2$ clearly vanishes at $x = 0$. Since $W(x,x^2)$ vanishes at $x = 0$, but x and x^2 are linearly independent on $-1 < x < 1$, it follows that x and x^2 cannot be solutions of Eq.(6) on $-1 < x < 1$. To show that the functions $y_1 = x$ and $y_2 = x^2$ are solutions of $x^2y'' - 2xy' + 2y = 0$, substitute each of them in the equation. Clearly they are solutions. There is no contradiction to Theorem 3.5 since $p(x) = -2/x$ and $q(x) = 2/x^2$ are discontinuous at $x = 0$, and hence the theorem does not apply on the interval $-1 < x < 1$.

13. On $0 < x < 1$, $y_1(x) = x^2$ and $y_2(x) = x^2$. Hence there are nonzero constants, $c_1 = 1$ and $c_2 = -1$, such that $c_1y_1(x) + c_2y_2(x) = 0$ for each x in $(0,1)$. On $-1 < x < 0$, $y_1(x) = -x^2$ and $y_2(x) = x^2$; thus $c_1 = c_2 = 1$ defines constants such that $c_1y_1(x) + c_2y_2(x) = 0$ for each x in $(-1,0)$. Thus y_1 and y_2 are linearly dependent on $0 < x < 1$ and on $-1 < x < 0$. We will show that $y_1(x)$ and $y_2(x)$ are linearly independent on $-1 < x < 1$ by demonstrating that it is impossible to find constants c_1 and c_2, not both zero, such that $c_1y_1(x) + c_2y_2(x) = 0$ for all x in $(-1,1)$. Assume that there are two such nonzero constants and choose two points x_0 and x_1 in $-1 < x < 1$ such that $x_0 < 0$ and $x_1 > 0$. Then $-c_1x_0^2 + c_2x_0^2 = 0$ and $c_1x_1^2 + c_2x_1^2 = 0$. These equations have a nontrivial solution for c_1 and

c_2 only if the determinant of coefficients is zero. But the determinant of coefficients is $-2x_0^2 x_1^2 \neq 0$ for x_0 and x_1 as specified. Hence $y_1(x)$ and $y_2(x)$ are linearly independent on $-1 < x < 1$.

14. Suppose that $x = a$ and $x = b$ ($b > a$) are consecutive zeros of y_1. We must show that y_2 vanishes once and only once in the interval $a < x < b$. Assume that it does not vanish. Then we can form the quotient y_1/y_2 on the interval $a \le x \le b$. Note $y_2(a) \neq 0$ and $y_2(b) \neq 0$, otherwise y_1 and y_2 would not be linearly independent solutions. Next y_1/y_2 vanishes at $x = a$ and $x = b$ and has a derivative in $a < x < b$. By Rolles theorem, the derivatives must vanish at an interior point. But

$$\left(\frac{y_1}{y_2}\right)' = \frac{y_1' y_2 - y_2' y_1}{y_2^2} = \frac{-W(y_1, y_2)}{y_2^2}$$

which cannot be zero since y_1 and y_2 are linearly independent solutions. Hence we have a contradiction, and we conclude that y_2 must vanish at a point between a and b. Finally, we show that it can vanish at only one point between a and b. Suppose that it vanishes at two points c and d between a and b. By the argument we have just given we can show that y_1 must vanish between c and d. But this contradicts the hypothesis that a and b are consecutive zeros of y_1.

Section 3.4, Page 130

3. Proceed as in Example 1. Let $y_2 = 1 \cdot v = v$. Then $y_2' =$
    v', $y_2'' = v''$ and on substituting in the D.E. we obtain
    $x^2 v'' + 2xv' = (x^2 v')' = 0$. Hence $v' = c_1 x^{-2}$ and $v =$
    $-c_1 x^{-1} + c_2$. Thus a second solution is $y_2(x) = 1 \cdot$
    $(-c_1 x^{-1} + c_2)$. The constant c_2 adds only a multiple of
    the constant solution y_1, so we can ignore it. We can
    also take $c_1 = -1$. Two linearly independent solutions
    are $y_1(x) = 1$ and $y_2(x) = x^{-1}$. The solutions are valid
    for $x > 0$ or $x < 0$. Note that when the D.E. is put into
    the standard form the coefficient of y' is discontinuous
    at $x = 0$.

5. Let $y_2 = v/x$. Then $y_2' = v'/x - v/x^2$ and $y_2'' = v''/x -$
    $2v'/x^2 + 2v/x^3$. Substituting in the D.E. we obtain
    $x^2(v''/x - 2v'/x^2 + 2v/x^3) + 3x(v'/x - v/x^2) + v/x = 0$.
    Simplifying the left side we get $xv'' + v' = 0$, which
    yields $v' = c_1/x$. Thus $v = c_1 \ln x + c_2$. Hence a second
    solution is $y_2(x) = (c_1 \ln x + c_2)/x$. However, we may set
    $c_2 = 0$ and $c_1 = 1$ without loss of generality and thus we
    have $y_2(x) = (\ln x)/x$ as a second solution. Note that in
    the form we actually calculated, $y_2(x)$ is a linear
    combination of $1/x$ and $\ln x/x$, and hence is the general
    solution.

8. In this case the calculations are somewhat easier if we

do not use the explicit form for $y_1(x) = \sin x^2$ at the

beginning but simply set $y_2(x) = y_1 v$. Substituting this

form for y_2 in the D.E. gives $x(y_1 v)'' - (y_1 v)' + 4x^3(y_1 v)$

$= 0$. On carrying out the differentiations and making use

of the fact that y_1 is a solution, we obtain $xy_1 v'' +$

$(2xy_1' - y_1)v' = 0$. This is a first order linear equation

for v', which has the solution $v' = cx/(\sin x^2)^2$.

Setting $u = x^2$ allows integration of this to get $v =$

$c_1 \cot x^2 + c_2$. Setting $c_1 = 1$, $c_2 = 0$ and multiplying

by $y_1 = \sin x^2$ we obtain $y_2(x) = \cos x^2$ as the second

solution of the D.E.

12. That $y_1(x) = x^{-1/2} \sin x$ is a solution of the D.E. is

verified by direct substitution. To obtain a second

solution set $y_2(x) = y_1(x)v(x)$. Substituting y_2 in the

D.E. gives $x^2(y_1 v)'' + x(y_1 v)' + (x^2 - \frac{1}{4})y_1 v = 0$. On

carrying out the differentiations and making use of the

fact that y_1 is a solution, we obtain $x^2 y_1 v'' + (2x^2 y_1' +$

$xy_1)v' = 0$. This is a first order linear equation for

v', $v'' + (2y_1'/y_1 + 1/x)v' = 0$ with solution

$$v'(x) = c \exp[- \int(2 \frac{y_1'}{y_1} + \frac{1}{x})dx] = c \exp[- 2 \ln y_1 - \ln x]$$

$$= c \frac{1}{xy_1^2} = \frac{c}{x(x^{-1} \sin^2 x)} = c \csc^2 x,$$

where c is an arbitrary constant, which we will take to

be one. Then $v(x) = \int \csc^2 x \, dx = - \cot x + k$ where again

k is an arbitrary constant which can be taken equal to

zero. Thus $y_2(x) = y_1(x)v(x) = (x^{-1/2} \sin x)(- \cot x) =$

$- x^{-1/2} \cos x$. The minus sign is unimportant since the

solution of a homogeneous equation can be multiplied by

any constant. The second solution is usually taken to be

$x^{-1/2} \cos x$. Note that $c = -1$ would have given this

solution.

13a. That $y_1(x) = e^x$ is a solution of the D.E. is verified by

direct substitution.

13b. Let $y_2(x) = e^x v(x)$, then $y_2' = e^x v' + e^x v$, and $y_2'' = e^x v'' +$

$2e^x v' + e^x v$. Substituting in the D.E. we obtain $xe^x v'' +$

$(xe^x - Ne^x)v' = 0$, or $v'' + (1 - N/x)v' = 0$. This is a first

order linear D.E. for v' with integrating factor $\mu(x) =$

$\exp[\int (1 - N/x)dx] = x^{-N} e^x$. Hence $(x^{-N} e^x v')' = 0$, and $v' =$

$c \, x^N e^{-x}$ which gives $v(x) = c \int x^N e^{-x} dx + k$. On taking $k =$

0 we obtain as the second solution $y_2(x) = ce^x \int x^N e^{-x} dx$.

The integral can be evaluated by using the method of

integration by parts. At each stage let $u = x^N$ or x^{N-1},

or whatever the power of x that remains, and let $dv =$

e^{-x}. Note that this dv is not related to the $v(x)$ in

$y_2(x)$. For $N = 2$ we have

$$y_2(x) = ce^x \int x^2 e^{-x} dx = ce^x [x^2 \frac{e^{-x}}{-1} - \int 2x \frac{e^{-x}}{-1} \, dx]$$

$$= - cx^2 + ce^x[2x \frac{e^{-x}}{-1} - \int 2 \frac{e^{-x}}{-1} dx]$$

$$= c(-x^2 - 2x - 2) = - 2c(1 + x + x^2/2!).$$

Choosing $c = -1/2!$ gives the desired result. For the general case $c = - 1/N!$.

15. $(y_2/y_1)' = (y_1 y_2' - y_1' y_2)/y_1^2 = W(y_1, y_2)/y_1^2$. Abel's identity is $W(y_1, y_2) = c \exp[-\int^x p(t)dt]$. Hence $(y_2/y_1)'$ $= cy_1^{-2} \exp[-\int^x p(t)dt]$. Integrating and setting $c = 1$ (since a solution y_2 can be multiplied by any constant) and the constant of integration equal to zero we obtain

$$y_2(x) = y_1(x) \int^x \frac{1}{[y_1(s)]^2} \exp[-\int^s p(t)dt]ds.$$

16. Dividing by x^2 we obtain $p(x) = 2/x$. Abel's identity gives $W(y_1, y_2)(x) = c \exp[-\int^x 2t^{-1} dt] = cx^{-2}$. From the result of Problem 15 we have $(y_2/y_1)' = W(y_1, y_2)/y_1^2$ or $(y_2/x)' = cx^{-2}/x^2 = cx^{-4}$. Hence $y_2(x) = - c/3x^2 + kx$. Choosing $c = -3$ and $k = 0$ (kx is simply a multiple of the solution y_1), we obtain $y_2(x) = x^{-2}$.

Section 3.5, Page 136

1. Assume $y = e^{rx}$, which is substituted into the D.E. to obtain the characteristic equation $r^2 + 2r - 3 = 0$. Hence $r_1 = 1$ and $r_2 = -3$. The general solution is then $y = c_1 e^x + c_2 e^{-3x}$.

6. Substituting $y = e^{rx}$ into the D.E., we find that $r^2 - 2r +$

$1 = 0$, which gives $r_1 = 1$ and $r_2 = 1$. Since the roots are equal, the second linearly independent solution is xe^x and thus the general solution is $y = c_1e^x + c_2 xe^x$.

13. Substituting $y = e^{rx}$ we find that $r^2 + 8r - 9 = 0$. The roots are $r_1 = 1$ and $r_2 = -9$, and the general solution is $y = c_1e^x + c_2e^{-9x}$. Since the I.C. are given at $x = 1$, it is convenient to introduce the arbitrary constants c_1 and c_2 in a slightly different way. We write the general solution in the form $y = k_1e^{(x-1)} + k_2e^{-9(x-1)}$. Note that $c_1 = k_1e^{-1}$ and $c_2 = k_2e^9$. The advantage of the latter form of the general solution becomes clear when we apply the I.C. $y(1) = 1$ and $y'(1) = 0$. This latter form of y gives $y' = k_1e^{(x-1)} - 9k_2e^{-9(x-1)}$ and thus setting $x = 1$ in y and y' yields the equations $k_1 + k_2 = 1$ and $k_1 - 9k_2 = 0$. Solving for k_1 and k_2 we find that $y = (9e^{(x-1)} + e^{-9(x-1)})/10$.

14. Show that the general solution can be written in the form $y = k_1e^{-2(x+1)} + k_2xe^{-2(x+1)}$.

17. If $r_2 \neq r_1$ then $\phi(x; r_1, r_2) = (e^{r_2x} - e^{r_1x})/(r_2 - r_1)$ is defined for all x. Note that ϕ is a linear combination of two solutions, e^{r_1x} and e^{r_2x}, of the D.E. Hence, ϕ is a solution of the differential equation. Think of r_1 as fixed and let $r_2 \to r_1$. The limit of ϕ as $r_2 \to r_1$ is indeterminate. If we use L'Hopital's rule, we find

$$\lim_{r_2 \to r_1} \frac{e^{r_2 x} - e^{r_1 x}}{r_2 - r_1} = \lim_{r_2 \to r_1} \frac{xe^{r_2 x}}{1} = xe^{r_1 x}$$

Hence, the solution $\phi(x;r_1,r_2) \to xe^{r_1 x}$ as $r_2 \to r_1$.

Section 3.5.1, Page 140

1. As in Section 3.5, we seek solutions of the form $y = e^{rx}$.
    Substituting this into the D.E. yields the characteristic
    equation $r^2 - 2r + 2 = 0$, which has the roots $r_1 = 1 + i$
    and $r_2 = 1 - i$, using the quadratic formula. Thus $\lambda = 1$
    and $\mu = 1$ and from Eq.(10) the general solution is $y = c_1 e^x \cos x + c_2 e^x \sin x$.

8. Assuming $y = e^{rx}$ yields the characteristic equation $r^2 + 4r + 5 = 0$, which has the roots $r_1, r_2 = -2 \pm i$. Thus
    the general solution is $y = c_1 e^{-2x} \cos x + c_2 e^{-2x} \sin x$.
    This gives $y' = -2c_1 e^{-2x} \cos x - c_1 e^{-2x} \sin x - 2c_2 e^{-2x} \sin x + c_2 e^{-2x} \cos x$. Thus $y(0) = c_1 + 0 = 1$ and
    $y'(0) = -2c_1 - 0 - 0 + c_2 = 0$, which yield $c_1 = 1$ and $c_2 = 2$. Hence $y = e^{-2x} \cos x + 2e^{-2x} \sin x$.

12. The general solution of the D.E. is $y = c_1 e^{r_1 x} + c_2 e^{r_2 x}$
    where $r_1, r_2 = (-b \pm \sqrt{b^2 - 4ac})/2a$ provided $b^2 - 4ac \neq 0$.
    In this case there are two possibilities. If $b^2 - 4ac > 0$
    then $(b^2 - 4ac)^{1/2} < b$ and r_1 and r_2 are real and
    <u>negative</u>. Consequently $e^{r_1 x} \to 0$ and $e^{r_2 x} \to 0$; and hence
    $y \to 0$, as $x \to \infty$. If $b^2 - 4ac < 0$ then r_1 and r_2 are

complex conjugates with <u>negative</u> real part. Again $e^{r_1 x} \to 0$

and $e^{r_2 x} \to 0$; and hence $y \to 0$, as $x \to \infty$. Finally, if

$b^2 - 4ac = 0$, then $y = c_1 e^{r_1 x} + c_2 x e^{r_1 x}$ where $r_1 = -b/2a < 0$

Hence, again $y \to 0$ as $x \to \infty$. This conclusion does not

hold if either $b = 0$ or $c = 0$.

15. $Ae^{i\theta} = A(\cos\theta + i \sin\theta)$. Equating real and imaginary

parts of $a + ib = Ae^{i\theta} = A \cos\theta + i A \sin\theta$, we obtain a

$= A \cos\theta$ and $b = A \sin\theta$. Hence $A = (a^2 + b^2)^{1/2} > 0$,

with θ determined by the equations $\cos\theta = a/A$, $\sin\theta =$

b/A, $0 < \theta < 2\pi$. Note that θ is only determined up to an

additive multiple of 2π; that is if θ is a solution of

the two equations then $\theta + 2m\pi$, $m = \pm 1, \pm 2, \ldots$, is

also a solution.

16. We use the result of Problem 15 that $a + ib = Ae^{i\theta}$, where

$A = (a^2 + b^2)^{1/2}$ and $\cos\theta = a/A$, $\sin\theta = b/A$. For $1 + i$

we find $A = \sqrt{2}$ and $\theta = \pi/4$; hence the square roots of

$(1 + i)$ are $\pm 2^{1/4} e^{i\pi/8} = \pm 2^{1/4}(\cos \pi/8 + i \sin \pi/8)$.

17a. Assume a solution of the form $y = e^{rx}$. The

characteristic equation is $r^2 + ir + 2 = 0$. This

equation can be factored by inspection or we can use the

quadratic formula to obtain the roots. We have $r_1, r_2 =$

$[-i \pm \sqrt{(-1) - 8}]/2 = (-i \pm 3i)/2 = i$ and $-2i$. Hence,

the general solution is $y = c_1 e^{ix} + c_2 e^{-i2x}$. Note that

since the D.E. has complex-valued coefficients a

real-valued function, such as the real or imaginary part

of the solutions e^{ix} and e^{-i2x}, cannot satisfy the D.E.

This can be verified by direct substitution.

9a. We use the result of Problem 18. Note that $q(x) = e^{-x^2} > 0$

for $-\infty < x < \infty$. Next, we find that $(q' + 2pq)/q^{3/2} =$

0. Hence the D.E. can be transformed into an equation

with constant coefficients by letting $z = u(x) =$

$\int^x e^{-t^2/2} dt$. Substituting $z = u(x)$ in the differential

equation found in part (b) of Problem 18 we obtain, after

dividing by the coefficient of d^2y/dz^2, the D.E. d^2y/dz^2

$- y = 0$. Hence the general solution of the original D.E.

is $y = c_1 \cos z + c_2 \sin z$, $z = \int^x e^{-t^2/2} dt$.

20. Rewrite the D.E. as $y" + (\alpha/x)y' + (\beta/x^2)y = 0$ so that p

$= \alpha/x$ and $q = \beta/x^2$, which satisfy the condition of part

(c) of Problem 18. Thus $z = \int^x (1/t^2)^{1/2} dt = \ln x$ will

transform the D.E. into $dy^2/dz^2 + (\alpha-1) dy/dz + \beta y = 0$.

Note that since β is constant, it can be neglected in

defining z.

21a. By direct substitution or from Problem 20, $z = \ln x$ will

transform the D.E. into $d^2y/dz^2 + y = 0$, since $\alpha = 1$ and

$\beta = 1$. Thus $y = c_1 \cos z + c_2 \sin z$, with $z = \ln x$, $x > 0$.

22a. Solving the D.E. $y'/y = -u$, we obtain $y(x) = cE(x)$, where

$E(x) = \exp[- \int^x u(t)dt]$ and c is an arbitrary constant.

Then $y'(x) = - cu(x)E(x)$ and $y''(x) = -cu'(x)E(x) +$

$cu^2(x)E(x)$. Substituting in the D.E. for y and dividing

by $cE(x)$ gives $-u'(x) + u^2(x) + p(x)[-u(x)] + q(x) = 0$,

so $u' = q(x) - p(x)u + u^2$.

22b. Let $w = -y'/yq_2$. If we calculate w' and substitute for w

and w' in the general Riccati equation we obtain $-y''/yq_2$

$+ (y')^2/y^2q_2 + y'q_2'/yq_2^2 = q_0 + q_1 (-y'/yq_2) +$

$q_2(y')^2/(yq_2)^2$. On multiplying by yq_2^2 and simplifying

we obtain the stated result.

Section 3.6.1, Page 154

1. First we find the solution of the homogeneous D.E.

    Substituting $y = e^{rx}$ gives the characteristic equation r^2

    $+ r - 2 = (r+2)(r-1) = 0$. Hence $y_c = c_1e^{-2x} + c_2e^x$. For

    the particular solution we assume $y_p = Ax + B$. Since

    neither Ax or B are solutions of the homogeneous equation

    it is not necessary to modify the original assumption.

    Substituting y_p in the D.E. we obtain $0 + A -2(Ax+B) = 2x$

    or $-2A = 2$ and $A-2B = 0$. Solving for A and B we obtain

    $y = c_1e^{-2x} + c_2e^x - x - 1/2$ as the general solution.

    $y(0) = 0 \to c_1 + c_2 - 1/2 = 0$ and $y'(0) = 1 \to -2c_1 + c_2$

    $- 1 = 1$, which yield $c_1 = -1/2$ and $c_2 = 1$. Thus $y = e^x -$

    $(1/2)e^{-2x} - x - 1/2$.

4. Initially we assume $y_p = A + B \sin 2x + C \cos 2x$.

    However, since a constant is a solution of the related

homogeneous D.E. we must modify y_p by multiplying the

constant A by x and thus the correct form is y_p = Ax +

B sin 2x + C cos 2x.

6. First, solve the homogeneous D.E. Substituting $y = e^{rx}$

gives $r^2 - 2r + 1 = (r-1)^2 = 0$. Hence $y_c(x) = c_1 e^x +$

$c_2 x e^x$. Next assume the form for $y_p(x)$: $y_p(x) = $ (Ax +

B)e^x + C. However, since e^x and xe^x are solutions of the

homogeneous equation it is necessary to multiply the

first term by x^2. The correct form for $y_p(x)$ is $y_p(x) = $

(Ax^3 + Bx^2)e^x + C. Substituting in the D.E., we obtain

e^x[Ax^3 + (6A + B)x^2 + (6A + 4B)x + 2B] − 2e^x[Ax^3 + (3A +

B)x^2 + 2Bx] + e^x(Ax^3 + Bx^2) + C = xe^x + 4. This reduces

to e^x (6Ax + 2B) + C = xe^x + 4. Hence C = 4, B = 0, and

A = 1/6. The general solution of the D.E. is $y = c_1 e^x +$

$c_2 x e^x$ + (1/6)x$^3 e^x$ + 4. The I.C. y(0) = 1 → c_1 + 4 = 1

and y'(0) = 1 → $c_1 + c_2$ = 1. Hence c_1 = −3 and c_2 = 4.

8. The assumed form for y_p is: y_p = (Ax + B) sin 2x + (Cx +

D) cos 2x, which is appropriate for both terms appearing

on the right side of the D.E. Since none of the terms

appearing in y_p are solutions of the homogeneous

equation, we do not need to modify y_p.

13. On substituting $y = e^{rx}$ in the homogeneous D.E., we find

$y_c(x) = e^{-x/2}[c_1 \cos (\sqrt{3} \, x/2) + c_2 \sin (\sqrt{3} \, x/2)]$. The

nonhomogeneous term is not in the class of functions

considered in this section. However, it can be put in a

suitable form by using the trigonometric identity $\sin^2 x =$

$(1 - \cos 2x)/2$. Now we assume $y_p(x) = A \cos 2x + B \sin 2x$

+ C. For algebraic simplicity, find y_p for $y" + y' +$

$y = 1/2$, and then $y_p(x)$ for $y" + y' + y = -(\cos 2x)/2$ and

apply the principle of superposition. The first equation

gives $C = 1/2$. Substituting $A \cos 2x + B \sin 2x$ in the

second equation, and collecting terms gives $(- 3A +$

$2B)\cos 2x + (- 2A - 3B)\sin 2x = -(\cos 2x)/2$. Solving for

A and B gives $A = 3/26$ and $B = -1/13$. The general

solution is $y = e^{-x/2}[c_1 \cos (\sqrt{3} x/2) + c_2 \sin(\sqrt{3} x/2)] +$

$1/2 + (3/26) \cos 2x - (1/13)\sin 2x$.

14. First solve the homogeneous D.E. Substituting $y = e^{rx}$

gives $r^2 + r + 4 = 0$. Hence $y_c = e^{-x/2}[c_1 \cos(\sqrt{15} x/2) +$

$c_2 \sin(\sqrt{15} x/2)]$. We replace $\sinh x$ by $(e^x - e^{-x})/2$ and

then assume $y_p(x) = Ae^x + Be^{-x}$. Since neither e^x nor e^{-x}

are solutions of the homogeneous equation, there is no

need to modify our assumption for y_p. Substituting in the

D.E., we obtain $6Ae^x + 4Be^{-x} = e^x - e^{-x}$. Hence, $A = 1/6$

and $B = -1/4$. The general solution is $y = e^{-x/2}[c_1 \cdot$

$\cos (\sqrt{15} x/2) + c_2 \sin(\sqrt{15} x/2)] + e^x/6 - e^{-x}/4$. [For

this problem we could also have found a particular

solution as a linear combination of $\sinh x$ and $\cosh x$:

$y_p(x) = A \cosh x + B \sinh x$. Substituting this in the

D.E. gives $(5A + B) \cosh x + (A + 5B)\sinh x = 2 \sinh x$.

The solution is $A = -1/12$ and $B = 5/12$. A simple

calculation shows that $-(1/12)\cosh x + (5/12) \sinh x = e^x/6 - e^{-x}/4$.]

16. The solution of the homogeneous D.E. is $y_c = c_1 e^{-3x} + c_2$.

After inspection of the nonhomogeneous term, we assume

$y_p(x) = (A_0 x^4 + A_1 x^3 + A_2 x^2 + A_3 x + A_4) + (B_0 x^2 + B_1 x + B_2)e^{-3x} + C \sin 3x + D \cos 3x$. However, since e^{-3x} and a

constant are solutions of the homogeneous D.E., we must

multiply the coefficient of e^{-3x} and the polynomial by x.

The correct form for $y_p(x)$ is $y_p(x) = x(A_0 x^4 + A_1 x^3 + A_2 x^2 + A_3 x + A_4) + x(B_0 x^2 + B_1 x + B_2)e^{-3x} + C \sin 3x + D \cos 3x$.

19. The solution of the homogeneous D.E. is $y_c = e^{-x}[c_1 \cos x + c_2 \sin x]$. After inspection of the nonhomogeneous

term, we assume $y_p(x) = Ae^{-x} + (B_0 x^2 + B_1 x + B_2)e^{-x} \cos x + (C_0 x^2 + C_1 x + C_2)e^{-x} \sin x$. Since $e^{-x} \cos x$ and

$e^{-x} \sin x$ are solutions of the homogeneous D.E., it is

necessary to multiply both of these terms by x. Hence,

the correct form for $y_p(x)$ is $y_p(x) = Ae^{-x} + x(B_0 x^2 + B_1 x + B_2)e^{-x} \cos x + x(C_0 x^2 + C_1 x + C_2)e^{-x} \sin x$.

24. First solve the I.V.P. $y'' + y = t$, $y(0) = 0$, $y'(0) = 1$

for $0 \le t \le \pi$. The solution of the homogeneous D.E. is

$y_c(t) = c_1 \cos t + c_2 \sin t$. The correct form for $y_p(t)$ is $y_p(t) = A_0 t + A_1$. Substituting in the D.E. we find $A_0 = 1$ and $A_1 = 0$. Hence, $y = c_1 \cos t + c_2 \sin t + t$. Applying the I.C., we obtain $y = t$. For $t > \pi$ the complementary solution is $y_c(t) = D_1 \cos t + D_2 \sin t$ and the form for $y_p(t)$ is $y_p(t) = Ee^{\pi-t}$. Substituting $y_p(t)$ in the D.E., we obtain $Ee^{\pi-t} + Ee^{\pi-t} = \pi e^{\pi-t}$ so $E = \pi/2$. Hence the general solution for $t > \pi$ is $y = D_1 \cos t + D_2 \cdot \sin t + (\pi/2)e^{\pi-t}$. If y and y' are to be continuous at $t = \pi$, then the solutions and their derivatives for $t \leq \pi$ and $t > \pi$ must have the same value at $t = \pi$. These conditions require $\pi = -D_1 + \pi/2$ and $1 = -D_2 - \pi/2$. Hence $D_1 = -\pi/2$, $D_2 = -(1 + \pi/2)$, and

$$y = \phi(t) = \begin{cases} t, & 0 \leq t \leq \pi \\ -(\pi/2)\cos t - (1 + \pi/2)\sin t + (\pi/2)e^{\pi-t}, & t > \pi. \end{cases}$$

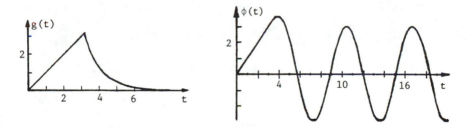

25. According to Theorem 3.11, the difference of any two solutions of the linear second order nonhomogeneous D.E. is a solution of the corresponding homogeneous D.E. Hence $y_1 - y_2$ is a solution of $ay'' + by' + cy = 0$. In

Problem 12 of Section 3.5.1 we showed that if $a > 0$, $b > 0$,

and $c > 0$ then every solution of this D.E. goes to zero as

$t \to \infty$.

28 a. From Problem 27 we write the D.E. as $(D-4)(D+1)y = 3e^{2x}$.

Thus let $(D+1)y = u$ and then $(D-4)u = 3e^{2x}$. This last

equation is the same as $du/dx - 4u = 3e^{2x}$, which may be

solved by multiplying both sides by e^{-4x} and integrating

(see section 2.1). This yields $u = (-3/2)e^{2x} + Ce^{4x}$.

Substituting this form of u into $(D+1)y = u$ we obtain

$dy/dx + y = (-3/2)e^{2x} + Ce^{4x}$. Again, multiplying by e^{x}

and integrating gives $y = (-1/2)e^{2x} + C_1e^{4x} + C_2e^{-x}$,

where $C_1 = C/5$.

Section 3.6.2, Page 160

2. Two linearly independent solutions of the homogeneous

D.E. are $y_1(x) = e^{2x}$ and $y_2(x) = e^{-x}$. Assume $y_p(x) =$

$u_1(x)e^{2x} + u_2(x)e^{-x}$, then $y_p'(x) = [2u_1(x)e^{2x} - u_2(x)e^{-x}]$

$+ [u_1'(x)e^{2x} + u_2'(x)e^{-x}]$. We set $u_1'(x)e^{2x} + u_2'(x)e^{-x} = 0$.

Computing $y_p''(x)$ and substituting in the D.E. gives

$2u_1'(x)e^{2x} - u_2'(x)e^{-x} = 2e^{-x}$. Thus we have two algebraic

equations for $u_1'(x)$ and $u_2'(x)$ with the solution $u_1'(x) =$

$2e^{-3x}/3$ and $u_2'(x) = -2/3$. Hence $u_1(x) = -2e^{-3x}/9$ and

$u_2(x) = -2x/3$. Substituting in the formula for $y_p(x)$ we

obtain $y_p(x) = (-2e^{-3x}/9)e^{2x} + (-2x/3)e^{-x} = (-2e^{-x}/9) -$

$(2xe^{-x}/3)$. Since e^{-x} is a solution of the homogeneous

D.E., we can choose $y_p(x) = -2xe^{-x}/3$.

4.   Since cos x and sin x are solutions of the homogeneous

D.E., we assume $y_p = u_1(x) \cos x + u_2(x) \sin x$.

Thus $y_p' = - u_1(x) \sin x + u_2(x) \cos x$ after we set $u_1'(x)$

· cos x + $u_2'(x) \sin x = 0$. Finding y_p'' and substituting

into the D.E. then yields $- u_1'(x) \sin x + u_2'(x) \cos x =$

tan x. The two equations for $u_1'(x)$ and $u_2'(x)$ have the

solutions: $u_1'(x) = - \sin^2 x/\cos x = - \sec x + \cos x$ and

$u_2'(x) = \sin x$. Thus $u_1(x) = \sin x - \ln(\tan x + \sec x)$

and $u_2(x) = - \cos x$, which when substituted into the

assumed form for y_p yields the answer.

8.   That x and xe^x are solutions of the homogeneous D.E. can

be verified by direction substitution. Thus we assume y_p

$= xu_1(x) + xe^x u_2(x)$. Following the pattern of Problems 2

and 4, we find $x u_1'(x) + xe^x u_2'(x) = 0$ and $u_1'(x) +$

$(x+1)e^x u_2' = 2x$. [Note that g(x) = 2x, since the D.E.

must be put into the form of Eq.(1)]. The solution of

these equations gives $u_1'(x) = -2$ and $u_2'(x) = 2e^{-x}$.

Hence, $u_1(x) = -2x$ and $u_2(x) = -2e^{-x}$, and $y_p(x) = x(-2x)$

$+ xe^x(-2e^{-x}) = -2x^2 - 2x$. However, since x is a solution

of the homogeneous D.E. we can choose as our particular

solution $y_p(x) = -2x^2$.

10. For this problem, and for many others, it is probably

easier to rederive Eqs.(8) without using the explicit

form for $y_1(x)$ and $y_2(x)$ and then to substitute for $y_1(x)$

and $y_2(x)$ in Eqs.(8). In this case if we take $y_1 = x^{-1/2}$

$\cdot \sin x$ and $y_2 = x^{-1/2} \cos x$, then $W(y_1,y_2) = -1/x$. If

the D.E. is put in the form of Eq.(1), then $g(x) = 3x^{-1/2}$

$\cdot \sin x$ and thus $u_1'(x) = 3 \sin x \cos x$ and $u_2'(x) =$

$-3 \sin^2 x = 3(-1 + \cos 2x)/2$. Hence $u_1(x) = (3 \sin^2 x)/2$

and $u_2(x) = -3x/2 + 3(\sin 2x)/4$, and

$$y_p(x) = \frac{3 \sin^2 x}{2} \frac{\sin x}{\sqrt{x}} + (-\frac{3x}{2} + \frac{3 \sin 2x}{4}) \frac{\cos x}{\sqrt{x}}$$

$$= \frac{3 \sin^2 x}{2} \frac{\sin x}{\sqrt{x}} + (-\frac{3x}{2} + \frac{3 \sin x \cos x}{2}) \frac{\cos x}{\sqrt{x}}$$

$$= \frac{3 \sin x}{2\sqrt{x}} - \frac{3\sqrt{x} \cos x}{2}.$$

The first term is a multiple of $y_1(x)$, so the general

solution is $y = c_1 x^{-1/2} \sin x + c_2 x^{-1/2} \cos x - \frac{3}{2} x^{1/2}$

$\cdot \cos x.$

12. Two linearly independent solutions of the homogeneous

D.E. are $y_1(x) = e^{3x}$ and $y_2(x) = e^{2x}$. Applying Theorem

3.13 with $W(y_1,y_2)(x) = -e^{5x}$, we obtain

$$y_p(x) = -e^{3x} \int^x \frac{e^{2t}g(t)}{-e^{5t}} dt + e^{2x} \int^x \frac{e^{3t}g(t)}{-e^{5t}} dt$$

$$= \int^x [e^{3(x-t)} - e^{2(x-t)}]g(t)\, dt.$$

14. Since $y_c = c_1 + c_2 x$ we assume $y_p = u_1(x) + xu_2(x)$.

Following the pattern of the previous problems, we find

$u_1'(x) = - xg(x)$ and $u_2'(x) = g(x)$ and thus $y_p = -\int^x t\ g(t)$

$\cdot dt + x \int^x g(t)\ dt$. The first integral may be integrated

by parts by letting $u = t$ and $dv = g(t)dt$. Thus $du = dt$

and $v = \int^x g(t)dt$. Hence $y_p = -x \int^x g(t)dt + \int^x (\int^s g(t)dt)ds$

$+ x \int^x g(t)dt = \int^x(\int^s g(t)dt)ds$. By choosing the lower

limits to be zero, we are effectively choosing the

constants of integration so that $y_p(0) = 0$ and $y_p'(0) = 0$.

[Note: y_p' is found to be: $y_p' = \int_o^x g(t)dt$ using the

Fundamental Theorem of Calculus]. The general solution

of the D.E. is then $y = c_1 + c_2\ x + \int_o^x (\int_o^s g(t)dt)ds$,

where c_1 and c_2 are found to be y_o and y_o' respectively

when the I.C. are satisfied.

15a. To differentiate $y_p(x)$ you must use Leibnitz's Rule,

since the variable x appears in the limits of integration

as well as under the integral.

18a. First, we put the D.E. in standard form by dividing by

x^2: $y'' - 2y'/x + 2y/x^2 = 4$. Assuming that $y = xv(x)$ and

substituting in the D.E. we obtain $xv'' = 4$. Hence $v'(x)$

$= 4\ \ln x + c_2$ and $v(x) = 4 \int^x \ln s\ ds + c_2 x = 4(x\ \ln x -$

$x) + c_2 x$. The general solution is $c_1 y_1(x) + xv(x) = c_1 x$

$+ 4(x^2\ \ln x - x^2) + c_2 x^2$. Since $-4x^2$ is a multiple of

$y_2 = c_2 x^2$ we can write $y = c_1 x + c_2 x^2 + 4x^2\ \ln x$.

Section 3.7.1, Page 171

2.   The motion is an undamped free vibration. The units are

in the CGS system. The spring constant k = (100 gm) (980

cm/sec^2)/5 cm. Hence the D.E. for the motion is $100\ddot{u}$ +

[(100 · 980)/5]u = 0 where u is measured in cm and time in

sec. We obtain $\ddot{u} + 196u = 0$ so u = A cos 14t + B sin 14t.

The I.C. are u(0) = 0 → A = 0 and $\dot{u}(0)$ = 10 cm/sec →

B = 10/14 = 5/7. Hence u(t) = (5/7)sin 14t.

3. This is an undamped free vibration. The spring constant

k = 3 1b/0.25 ft = 12 1b/ft, and the mass m = 3/32

1b-sec^2/ft. Hence the I.V.P. is $(3/32)\ddot{u} + 12u = 0$, u(0) =

-1/12, $\dot{u}(0)$ = 2 ft/sec. The solution is u =

$(\sqrt{2}/8)\sin(8\sqrt{2}t) - (1/12)\cos(8\sqrt{2}t)$ ft, which has the

amplitude $(2/64 + 1/144)^{1/2} = \sqrt{22}/24$ ft, circular

frequency $\omega_0 = (k/m)^{1/2} = (12 \cdot 32/3)^{1/2} = 8\sqrt{2}$ rad/sec

and period $T = 2\pi/\omega_0 = \pi\sqrt{2}/8$ sec.

5. For this problem m = (3/32) 1b-sec^2/ft, c = 0.3

1b-sec/ft, and k = 12 1b/ft. The D.E. is $(3/32)\ddot{u} + 0.3\dot{u}$

$+ 12u = 0$ or $3\ddot{u} + 9.6\dot{u} + 384u = 0$ where u is measured in

ft and t in sec. We look for solutions of the form u =

exp(rt), and obtain r = -1.6 $\pm$ i(11.2). Thus

the quasi-period is $T_d = (2\pi/\mu)$sec where μ = 11.2. The

natural period is $T = \pi\sqrt{2}/8$. Hence, the percentage

change based on the natural period is $100(T_d - T)/T \simeq 1\%$.

9. The tangential force is mg sin θ in the direction of

decreasing θ . The acceleration is $\ell\ddot{\theta}$ which is measured as positive in the increasing θ direction. Hence $m\ell\ddot{\theta} =$ $-mg \sin\theta$. For θ small, $\sin\theta \simeq \theta$ and we obtain the D.E. $\ell\ddot{\theta} + g\theta = 0$. The general solution of this D.E. is $\theta =$ $A \cos\sqrt{g/\ell}\ t + B \sin\sqrt{g/\ell}\ t$, where A and B are arbitrary constants. Hence, the natural period is $T = 2\pi/\sqrt{g/\ell} =$ $2\pi(\ell/g)^{1/2}$. Note that the period does not depend on m.

10. First, consider the static case. Let $\Delta\ell$ denote the length of the block below the surface of the water. The weight of the block, which is a downward force, is $w = \rho\ell^3 g$. This is balanced by an equal and opposite buoyancy force B, which is equal to the weight of the displaced water. Thus $B = (\hat{\rho}\ell^2\Delta\ell)g = \rho\ell^3 g$ so $\hat{\rho}\Delta\ell = \rho\ell$. Now let x be the displacement of the block from its equilibrium position. We take downward as the positve direction. In a displaced position the forces acting on the block are its weight, which acts downward and is unchanged, and the buoyancy force which is now $\hat{\rho}\ell^2$ $\cdot (\Delta\ell + x)g$ and acts upward. The resultant force must be equal to the mass of the block times the acceleration, namely $\rho\ell^3\ddot{x}$. Hence $\rho\ell^3 g - \hat{\rho}\ell^2(\Delta\ell + x)g = \rho\ell^3\ddot{x}$. The D.E. for the motion of the block is $\rho\ell^3\ddot{x} + \hat{\rho}\ell^2 gx = 0$. This gives a simple harmonic motion with frequency $(\hat{\rho}g/\rho\ell)^{1/2}$ and natural period $2\pi(\rho\ell/\hat{\rho}g)^{1/2}$.

11. The spring constant is k = (20)(980)/5 = 3920 dyne/cm.

The I.V.P. for the motion is $20\ddot{u} + 400\dot{u} + 3920u = 0$ or $\ddot{u}$

$+ 20\dot{u} + 196u = 0$ and $u(0) = 2$, $\dot{u}(0) = 0$. Here u is

measured in cm and t in sec. The general solution of the

D.E. is $u = Ae^{-10t} \cos 4\sqrt{6}\, t + Be^{-10t} \sin 4\sqrt{6}\, t$. The

I.C. $u(0) = 2 \rightarrow A = 2$ and $\dot{u}(0) = 0 \rightarrow -10A + 4\sqrt{6}\, B = 0$.

The solution is

$$u = e^{-10t}[2 \cos 4\sqrt{6}\, t + 5(\sin 4\sqrt{6}\, t)/\sqrt{6}]\text{cm}.$$

13. The mass is 8/32 lb-sec^2/ft, and the spring constant is

8/(1/8) = 64 lb/ft. Hence $(1/4)\ddot{u} + c\dot{u} + 64u = 0$ or $\ddot{u} +$

$4c\dot{u} + 256u = 0$, where u is measured in ft, t in sec and

the units of c are lb-sec/ft. We look for solutions of

the D.E. of the form $u = e^{rt}$ and find $r^2 + 4cr + 256$ so

$r_1, r_2 = [-4c \pm \sqrt{16c^2 - 1024}]/2$. The system will be

overdamped, critically damped or underdamped as $(16c^2 -$

1024) is > 0, = 0, or < 0, respectively. Thus the system

is overdamped when c > 8 lb-sec/ft, critically damped

when c = 8 lb-sec/ft, and underdamped when c < 8

lb-sec/ft.

16. The general solution of the D.E. is $u = Ae^{r_1 t} + Be^{r_2 t}$

where $r_1, r_2 = [- c \pm (c^2 - 4km)^{1/2}]/2m$ provided $c^2 - 4km$

$\neq 0$, and where A and B are determined by the I.C. When

the motion is overdamped, $c^2 - 4km > 0$ and $r_1 > r_2$.

Setting u = 0, we obtain $Ae^{r_1 t} = -Be^{r_2 t}$ or $e^{(r_1-r_2)t} =$
$-B/A$. Since the exponential function is a monotone
function, there is at most one value of t (when $B/A < 0$)
for which this equation can be satisfied. Hence u can
vanish at most once. If the system is critically damped,
the general solution is $u(t) = (A + Bt)e^{-ct/2m}$. The
exponential function is never zero; hence u can vanish
only if $A + Bt = 0$. If $B = 0$ then u never vanishes; if
$B \neq 0$ then u vanishes once at $t = -A/B$ provided $A/B < 0$.

17. The general solution of Eq.(6) for the case of critical
damping is $u = (A + Bt)e^{-ct/2m}$. The I.C. $u(0) = u_0 \to A =$
u_0 and $\dot{u}(0) = \dot{u}_0 \to \dot{u}_0 = A(-c/2m) + B$. Hence $u = [u_0 +$
$(\dot{u}_0 + cu_0/2m)t]e^{-ct/2m}$. We now ask what conditions will
insure that u = 0 at least once. Since the exponential
function is never zero we require $u_0 + (\dot{u}_0 + cu_0/2m)t = 0$
at a positive value of t. This requires that $\dot{u}_0 + cu_0/2m$
$\neq 0$ and that $t = -u_0(\dot{u}_0 + cu_0/2m)^{-1} > 0$. We know that
$u_0 > 0$ so we must have $\dot{u}_0 + cu_0/2m < 0 \to \dot{u}_0 < -cu_0/2m$.

19. From Eq.(10), the quasi-circular frequency is given by u
$= (k/m - c^2/4m^2)^{1/2}$ and from Eq.(3) we have $\omega_0^2 = k/m$.
Hence $\mu = (\omega_0^2 - c^2/4m^2)^{1/2}$ and thus μ is always less
than ω_0.

Section 3.7.2, Page 178

1.   We use the trigonometric identities $\cos(A \pm B) = \cos A$
     $\cdot \cos B \mp \sin A \sin B$ to obtain a formula for the
     difference of two cosines in terms of sines. We have
     $\cos(A + B) - \cos(A - B) = -2 \sin A \sin B$. If we choose
     $A + B = 9t$ and $A - B = 7t$, then $A = 8t$ and $B = t$.
     Substituting in the formula just derived, we obtain
     $\cos 9t - \cos 7t = -2 \sin 8t \sin t$.

3.   The mass $m = 4/32 = 1/8$ lb-sec^2/ft and the spring
     constant $k = 4/(1/8) = 32$ lb/ft. Since there is no
     damping, the I.V.P. is $(1/8)\ddot{u} + 32u = 2 \cos 3t$, $u(0) =$
     $1/6$, $\dot{u}(0) = 0$ where u is measured in ft and t in sec.
     The natural frequency of the system is $\omega_0 = \sqrt{k/m} = 16$
     rad/sec. Notice that $\omega_0 \neq 3$; hence the general solution
     of the D.E. can be obtained by substituting in Eq.(3) or
     by using directly the method of undetermined
     coefficients: $u = c_1 \cos 16t + c_2 \sin 16t + (16/247)$
     $\cdot \cos 3t$. The I.C. $u(0) = 1/6 \rightarrow c_1 + 16/247 = 1/6$ and
     $\dot{u}(0) = 0 \rightarrow 16c_2 = 0$. The solution is $u = (151/1482)$
     $\cdot \cos 16t + (16/247)\cos 3t$ ft. Resonance occurs when the
     frequency ω of the forcing function $4 \sin \omega t$ is the same
     as the natural frequency of the system. When $\omega = 16$
     rad/sec the system will resonate.

4.   The I.V.P. is $(3/16)\ddot{u} + 12 u = 4 \cos 7t$, $u(0) = 0$, $\dot{u}(0) =$

O where u is measured in ft and t in sec. Since ω_0 = 8

and ω = 7, the particular solution has the form A cos 7t

+ B sin 7t. Using the method of undetermined

coefficients A = 64/45 and B = 0. Thus u = c_1 cos 8t +

c_2 sin 8t + (64/45) cos 7t. The I.C. yield c_1 = $-$ 64/45

and c_2 = 0, so that u = (64/45)(cos 7t $-$ cos 8t) =

(128/45)sin t/2 sin 15t/2 ft, using the trigonometic

identities of Problem 1. Since this I.V.P. is exactly

the same as discussed in the text, the solution can be

compared to Eq.(6). However, most I.V.P. will differ

from that of the text, and hence the steps used in

solving the problem here are important to know. The

graph of u is shown next, where the dashed curve

represents the time varying amplitude.

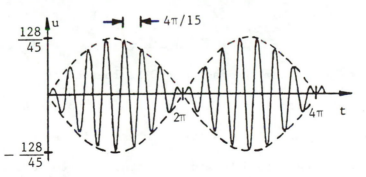

5.   Note that this problem involves resonance and thus u_p =

t(A cos 8t + B sin 8t).

7.   The D.E. is $m\ddot{u} + ku = F(t)$. If we divide by m and make

use of the fact that k/m = 1 we obtain $\ddot{u} + u = m^{-1}F(t)$.

We must solve the following D.E.: (1) $\ddot{u}_1 + u_1 = F_0 t/m$,

$0 < t < \pi$; (2) $\ddot{u}_2 + u_2 = F_0(2\pi - t)/m$, $\pi < t < 2\pi$; and (3)

$\ddot{u}_3 + u_3 = 0$, $t > 2\pi$. In addition we have the following

I.C. and matching conditions: $u_1(0) = \dot{u}_1(0) = 0$, $u_1(\pi) = u_2(\pi)$, $\dot{u}_1(\pi) = \dot{u}_2(\pi)$, $u_2(2\pi) = u_3(2\pi)$, $\dot{u}_2(2\pi) = \dot{u}_3(2\pi)$.

The conditions at π and 2π insure the continuity of u

and $\dot{u}$ at those points. The general solutions of the D.E.

are $u_1 = b_1 \cos t + b_2 \sin t + F_0 t/m$, $u_2 = c_1 \cos t + c_2$

$\cdot \sin t + F_0(2\pi - t)/m$, and $u_3 = d_1 \cos t + d_2 \sin t$. The

I.C. and matching conditions, in order, give $b_1 = 0$, b_2

$+ \pi F_0/m = 0$, $-b_1 + \pi F_0/m = -c_1 + F_0/m$, $-b_2 + F_0/m = -c_2$

$- F_0/m$, $c_1 = d_1$, and $c_2 - F_0/m = d_2$. Solving these

equations we obtain

$$u = (F_0/m) \begin{cases} t - \sin t & , \ 0 \leq t \leq \pi \\ (2 - t) - 3 \sin t, & \pi < t \leq 2\pi \\ - 4 \sin t & , \ 2\pi < t. \end{cases}$$

8. For this problem the mass $m = 8/32$ lb-sec^2/ft and the

spring constant $k = 8/(1/2) = 16$ lb/ft, so the D.E. is

$0.25\ddot{u} + 0.25\dot{u} + 16u = 4 \cos 2t$ where u is measured in ft

and t in sec. To determine the steady state response we

need only compute a particular solution of the

nonhomogeneous D.E. since the solutions of the

homogeneous D.E. decay to zero as $t \to \infty$. We assume

$u_p(t) = A \cos 2t + B \sin 2t$, and substitute in the D.E.:

$- A \cos 2t - B \sin 2t + (1/2)(- A \sin 2t + B \cos 2t) +$

$16(A \cos 2t + B \sin 2t) = 4 \cos 2t$. Hence $15A + (1/2)B =$

4 and $- (1/2)A + 15B = 0$, from which we obtain A =

240/901 and B = 8/901. The steady state response is

$u_p(t) = (240 \cos 2t + 8 \sin 2t)/901$.

9.   In order to determine the value of m that maximizes the

steady state response, we note that the present problem

has exactly the form of the problem considered in the

text, Eqs.(8) and (9). The response is a maximum when

$f(m) = m^2(\omega_0^2 - \omega^2)^2 + c^2\omega^2$, where $\omega_0^2 = k/m$, is a

minimum. We calculate df/dm, and set this quantity equal

to zero, and obtain $m = k/\omega^2$. We verify that this value

of m gives a minimum of $f(m)$ by the second derivative

test. For this problem $k = 16$ lb/ft and $\omega = 2$ rad/sec

so the value of m that maximizes the response of the

system is m = 4 slugs.

11. As suggested in the text, the maxima of $M(\omega)$ occur at the

minima of $\hat{m}(\omega) = m^2(\omega_0^2 - \omega^2)^2 + c^2\omega^2$.

Section 3.8, Page 182

2.   The D.E. for the system is $L\ddot{Q} + (1/C)Q = E_0 \cos \omega t$ or $\ddot{Q} +$

$\omega_0^2 Q = (E_0/L) \cos \omega t$ where $\omega_0^2 = 1/LC$. The general

solution of the homogeneous D.E. is $Q_c(t) = c_1 \cos \omega_0 t +$

$c_2 \sin \omega_0 t$. For $\omega \neq \omega_0$ the proper form for a particular solution is $Q_p(t) = A \cos \omega t + B \sin \omega t$. However, if $\omega = \omega_0$ these functions are solutions of the homogeneous D.E. and it is necessary to multiply our initial assumption by t. Thus if $\omega = \omega_0$, the correct form for Q_p is $Q_p(t) = At \cdot \cos \omega_0 t + Bt \sin \omega_0 t$. On substituting this expression in the D.E. we find $A = 0$ and $B = E_0/2\omega_0 L$ so the solution is $Q_p(t) = c_1 \cos \omega_0 t + c_2 \sin \omega_0 t + (E_0/2\omega_0 L)t \sin \omega_0 t$. Because of the last term the solution will become unbounded as $t \to \infty$. If damping is present, then neither $\cos \omega t$ nor $\sin \omega t$ can be a solution of the homogeneous D.E. and resonance cannot occur.

4a. Substituting the given values for L,C and R we obtain the D.E. $.2\ddot{Q} + 3 \times 10^2\, \dot{Q} + 10^5\, Q = 0$. The I.C. are $Q(0) = 10^{-6}$ and $\dot{Q}(0) = I(0) = 0$, since the circuit is originally open and thus there is no initial current. Assuming $Q = e^{rt}$, we obtain the roots of the characteristic equation as $r_1 = -500$ and $r_2 = -1000$. Thus $Q = c_1 e^{-500t} + c_2 e^{-1000t}$ and hence $Q(0) = 10^{-6} \to c_1 + c_2 = 10^{-6}$ and $\dot{Q}(0) = 0 \to -500c_1 - 1000c_2 = 0$. Solving for c_1 and c_2 yields the solution.

6. The I.V.P. is $\ddot{Q} + 5 \times 10^3\, \dot{Q} + 4 \times 10^6\, Q = 12$, $Q(0) = 0$, and $\dot{Q}(0) = 0$. The particular solution is of the form $Q_p = A$, so that upon substitution into the D.E. we obtain $4 \times 10^6 \cdot A = 12$ or $A = 3 \times 10^{-6}$. The general solution of the

D.E. is $Q = c_1 e^{r_1 t} + c_2 e^{r_2 t} + 3 \times 10^{-6}$, where r_1 and r_2 are found, as in Problem 4, to be $r_1 = -1000$ and $r_2 = -4000$. The I.C. yield $c_1 = -4 \times 10^{-6}$ and $c_2 = 10^{-6}$ and thus $Q = 10^{-6} (e^{-4000t} - 4e^{-1000t} + 3)$ coulombs.

Substituting $t = .001$ sec we obtain $Q(.001) = 10^{-6}(e^{-4} - 4e^{-1} + 3) = 10^{-6}(0.0183 - 4 \times 0.3679 + 3) = 1.55 \times 10^{-6}$ coulombs. Since the exponentials are to a negative power $Q(t) \to 3 \times 10^{-6}$ coulombs as $t \to \infty$, and this is the steady state charge.

7. The steady state current is given by Eq.(11). If this formula is not at hand, then it must be rederived by solving the D.E. $L\ddot{Q} + R\dot{Q} + (1/C)Q = E_0 \cos \omega_0 t$ and then calculating $I = \dot{Q}$. Note that it is probably easier to solve the D.E. in general and then substitute the explicit values for L, R, C, E_0 and ω than it is to solve the D.E. with the explicit values for the parameters. From Eq.(10), we have $I_p(t) = (\omega E_0/\Delta)\cos(\omega t + \delta)$ where (1) $\Delta^2 = (1/C - L\omega^2)^2 + \omega^2 R^2 = [4 \times 10^5 - 10 \times (120\pi)^2]^2 + (120\pi)^2 \times (9 \times 10^6) = 2.322 \times 10^{12}$ so $\Delta = 1.524 \times 10^6$; (2) $\sin \delta = (1/C - L\omega^2)/\Delta \simeq -1.021 \times 10^6/1.524 \times 10^6 = -0.6699$ and $\cos \delta = \omega R/\Delta \simeq 1.131 \times 10^6/1.524 \times 10^6 = 0.7421$ so $\delta \simeq \arctan(-0.90)$ and δ is in the fourth quadrant; and (3) $\omega E_0/\Delta \simeq 0.027$. Hence $I_p(t) = 0.027$

$\cdot \cos (120 t + \delta)$ amperes.

8. The steady state current is given by Eq.(11). $I_p(t)$ will

    be a maximum when $R^2 + [\omega L - 1/\omega c]^2$ is a minimum. Since

    this expression is the sum of two non-negative terms, one

    of which does not depend on ω, we choose ω to make the

    second term vanish: $\omega = (LC)^{-1/2}$.

CHAPTER 4

Section 4.1, Page 192

1b. Use the ratio test:

$$\lim_{n \to \infty} \frac{\left| (n+1)x^{n+1}/2^{n+1} \right|}{\left| nx^n/2^n \right|} = \lim_{n \to \infty} \frac{n+1}{n} \cdot \frac{1}{2} |x| = \frac{|x|}{2} .$$

Therefore the series converges absolutely for $|x| < 2$.
For $x = 2$ and at $x = -2$ the nth term does not approach
zero as $n \to \infty$ so the series diverge. Hence the radius
of convergence is $\rho = 2$.

1e. Use the ratio test:

$$\lim_{n \to \infty} \frac{\left| (2x+1)^{n+1}/(n+1)^2 \right|}{(2x+1)^n/n^2} = \lim_{n \to \infty} \frac{n^2}{(n+1)^2} |2x+1| = |2x+1| .$$

Therefore the series converges absolutely for
$|x+1/2| < 1/2$. At $x = 0$ and $x = -1$ the series also
converge absolutely. However, for $|x+1/2| > 1/2$ the
series diverges by the ratio test. The radius of
convergence is $\rho = 1/2$.

1h. Use the ratio test:

$$\lim_{n \to \infty} \frac{\left| (-1)^{n+1}[(n+1)!]^2 x^{2n+3}/(2n+2)! \right|}{\left| (-1)^n (n!)^2 x^{2n+1}/(2n)! \right|}$$

$$= \lim_{n \to \infty} \frac{2n!}{(2n+2)!} \frac{[(n+1)!]^2}{(n!)^2} x^2$$

$$= \lim_{n \to \infty} \frac{(n+1)^2}{(2n+2)(2n+1)} x^2 = \frac{1}{4} x^2 \ .$$

Thus the series converges absolutely for $x^2 < 4$, or

$|x| < 2$. Hence $\rho = 2$.

2a. For this problem $f(x) = \sin x$. Hence $f'(x) = \cos x$,

$f''(x) = - \sin x$, $f'''(x) = - \cos x, \ldots$. Then $f(0) = 0$,

$f'(0) = 1$, $f''(0) = 0$, $f'''(0) = -1, \ldots$. The even terms

in the series will vanish and the odd terms will

alternate in sign. We obtain

$\sin x = \sum\limits_{n=0}^{\infty} (-1)^n x^{2n+1}/(2n+1)!$. From the ratio test it
follows that $\rho = \infty$.

2d. For this problem $f(x) = x^2$. Hence $f'(x) = 2x$, $f''(x) = 2$,

and $f^{(n)}(x) = 0$ for $n > 2$. Then $f(-1) = 1$, $f'(-1) = -2$,

$f''(-1) = 2$ and $x^2 = 1 - 2(x+1) + 2(x+1)^2/2! = 1 - 2(x+1)$

$+ (x+1)^2$. Since the series terminates after a finite

number of terms, it converges for all x. Thus $\rho = \infty$.

2e. For this problem $f(x) = \ln x$. Hence $f'(x) = 1/x$, $f''(x)$

$= -1/x^2$, $f'''(x) = 1 \cdot 2/x^3, \ldots$, and $f^{(n)}(x) =$

$(-1)^{n+1}(n-1)!/x^n$. Then $f(1) = 0$, $f'(1) = 1$, $f''(1) = -1$,

$f'''(1) = 1 \cdot 2, \ldots$, $f^{(n)}(1) = (-1)^{n+1}(n-1)!$ The Taylor

series is $\ln x = (x-1) - (x-1)^2/2 + (x-1)^3/3 - \ldots =$

$\sum\limits_{n=1}^{\infty} (-1)^{n+1}(x-1)^n/n$. It follows from the ratio test that
the series converges absolutely for $|x-1| < 1$. However,

the series diverges at $x = 0$ so $\rho = 1$.

6.   If we shift the index of summation in the first sum by

letting m = n-1, we have

$\sum_{n=1}^{\infty} n \, a_n x^{n-1} = \sum_{m=0}^{\infty} (m+1) \, a_{m+1} x^m$. Substituting this

into the given equation and letting m = n again, we

obtain:

$$\sum_{n=0}^{\infty} (n+1)a_{n+1} x^n + 2 \sum_{n=0}^{\infty} a_n x^n = 0,$$

$$\sum_{n=0}^{\infty} [(n+1)a_{n+1} + 2a_n]x^n = 0 .$$

Hence $a_{n+1} = - 2a_n/(n+1)$ for n = 0,1,2,3,... . Thus $a_1 =$

$-2a_0$, $a_2 = -2a_1/2 = 2^2 a_0/2$, $a_3 = -2a_2/3 = -2^3 a_0/2 \cdot 3 =$

$-2^3 a_0/3!$... and $a_n = (-1)^n 2^n a_0/n!$. Notice that for n = 0

this formula reduces to a_0 so we can write $\sum_{n=0}^{\infty} a_n x^n =$

$\sum_{n=0}^{\infty} (-1)^n 2^n a_0 x^n/n! = a_0 \sum_{n=0}^{\infty} (-2x)^n/n! = a_0 e^{-2x}$.

Section 4.2, Page 206

2. $y = \sum_{n=0}^{\infty} a_n x^n$; $y' = \sum_{n=1}^{\infty} na_n x^{n-1}$ and since we must multiply

y' by x in the D.E. we do not shift the index; and

$y'' = \sum_{n=2}^{\infty} n(n-1)a_n x^{n-2} = \sum_{n=0}^{\infty} (n+2)(n+1)a_{n+2}x^n$.

Substituting in the D.E., we obtain

$$\sum_{n=0}^{\infty} (n+2)(n+1)a_{n+2}x^n - \sum_{n=1}^{\infty} na_n x^n - \sum_{n=0}^{\infty} a_n x^n = 0.$$

In order to have the starting point the same in all three

summations, we let n = 0 in the first and third terms to

obtain the following

$$(2 \cdot 1 \; a_2 - a_0)x^0 + \sum_{n=1}^{\infty} [(n+2)(n+1)a_{n+2} - (n+1)a_n]x^n = 0.$$

Thus $a_{n+2} = a_n/(n+2)$ for $n = 1,2,3,\dots$. However, note that recurrence relation is also correct for $n = 0$. The even and odd coefficients can be determined independently. We show how to calculate the odd a's:

$a_3 = a_0/3$, $a_5 = a_3/5 = a_0/5 \cdot 3$, $a_7 = a_5/7 = a_0/7 \cdot 5 \cdot 3, \dots$.
Now notice that $a_3 = 2a_0/(2 \cdot 3) = 2a_0/3!$, that $a_5 = 2 \; 4a_0/(2 \cdot 3 \cdot 4 \cdot 5) = 2^2 \cdot 2a_0/5!$, and that $a_7 = 2 \cdot 4 \cdot 6a_0/(2 \cdot 3 \cdot 4 \cdot 5 \cdot 6 \cdot 7) = 2^3 \cdot 3! \; a_0/7!$ Likewise $a_9 = a_7/9 = 2^3 \cdot 3! \; a_0/(7!)9 = 2^3 \cdot 3! \; 8a_0/9! = 2^4 \cdot 4! \; a_0/9!$. Continuing we have $a_{2m+1} = 2^m \cdot m! \; a_0/(2m+1)!$. In the same way we find that the even a's are given by $a_{2m} = a_0/2^m \; m!$. Thus

$$y = a_0 \sum_{m=0}^{\infty} \frac{x^{2m}}{2^m \; m!} + a_1 \sum_{m=0}^{\infty} \frac{2^m m! \; x^{2m+1}}{(2m+1)!} \; .$$

3. $y = \sum_{n=0}^{\infty} a_n(x-1)^n$; $y' = \sum_{n=1}^{\infty} na_n(x-1)^{n-1}$ and

$y'' = \sum_{n=2}^{\infty} n(n-1)a_n(x-1)^{n-2} = \sum_{n=0}^{\infty} (n+2)(n+1)a_{n+2}(x-1)^n.$
Substituting in the D.E. and replacing the coefficient x by $1 + (x-1)$ we obtain

$$\sum_{n=0}^{\infty} (n+2)(n+1)a_{n+2}(x-1)^n - \sum_{n=0}^{\infty} (n+1)a_{n+1}(x-1)^n - \sum_{n=1}^{\infty} na_n(x-1)^n$$

$$- \sum_{n=0}^{\infty} a_n(x-1)^n = 0,$$

where the index has been shifted in the second summation.

Letting n = 0 in the first, second and the fourth sums,

we obtain

$$(2 \cdot 1 \cdot a_2 - 1 \cdot a_1 - a_0)(x-1)^0 + \sum_{n=1}^{\infty} [(n+2)(n+1)a_{n+2}$$

$$- (n+1)a_{n+1} - (n+1)a_n](x-1)^n = 0.$$

Thus $(n+2)a_{n+2} - a_{n+1} - a_n = 0$ for $n = 0,1,2 \ldots$. This

recurrance relation can be used to solve for a_2 in terms

of a_0 and a_1, then for a_3 in terms of a_0 and a_1, etc. In

many cases it is easier to first take $a_0 = 0$ and generate

one solution and then take $a_1 = 0$ and generate the second

linearly independent solution. Thus, choosing $a_0 = 0$ we

find that $a_2 = a_1/2$, $a_3 = (a_2+a_1)/3 = a_1/2$, $a_4 =$

$(a_3+a_2)/4 = a_1/4$, $a_5 = (a_4+a_3)/5 = 3a_1/20,\ldots$. This

yields the solution $y_2(x) = a_1[(x-1) + (x-1)^2/2 +$

$(x-1)^3/2 + (x-1)^4/4 + 3(x-1)^5/20 + \ldots]$. The second

independent solution may be obtained by choosing $a_1 = 0$.

Then $a_2 = a_0/2$, $a_3 = (a_2+a_1)/3 = a_0/6$, $a_4 = (a_3+a_2)/4 =$

$a_0/6$, $a_5 = (a_4+a_3)/5 = a_0/15,\ldots$. This yields the

solution $y_1(x) = a_0[1 + (x-1)^2/2 + (x-1)^3/6 + (x-1)^4/6 +$

$(x-1)^5/15 + \ldots]$. Note that to obtain the answer given

in the text we can take $a_1 = 1$ in the y_2 solution and a_0

$= 1$ in the y_1 solution. The general solution is a linear

combination of y_1 and y_2.

5. $y = \sum\limits_{n=0}^{\infty} a_n x^n$; $y' = \sum\limits_{n=1}^{\infty} n a_n x^{n-1}$; and $y'' = \sum\limits_{n=2}^{\infty} n(n-1)a_n x^{n-2}$.

Substituting in the D.E. and shifting the index in the

series for y'' gives

$$\sum_{n=0}^{\infty} (n+2)(n+1)a_{n+2}x^n - \sum_{n=1}^{\infty} (n+1)n\ a_{n+1}x^n + \sum_{n=0}^{\infty} a_n x^n = 0,$$

$(2 \cdot 1 \cdot a_2 + a_0)x^0 + \sum\limits_{n=1}^{\infty} [(n+2)(n+1)a_{n+2} - (n+1)na_{n+1} +$

$a_n]x^n = 0$. Thus $a_2 = -a_0/2$ and $a_{n+2} = na_{n+1}/(n+2) -$

$a_n/(n+2)(n+1)$, $n = 1,2,\ldots$. Choosing $a_0 = 0$ yields $a_2 =$

0, $a_3 = -a_1/6$, $a_4 = 2\,a_3/4 = -a_1/12,\ldots$ which gives one

solution as $y_2(x) = a_1(x - x^3/6 - x^4/12 + \ldots)$. A

linearly independent solution is obtained by choosing a_1

$= 0$. Then $a_2 = -a_0/2$, $a_3 = a_2/3 = -a_0/6$, $a_4 = 2a_3/4 -$

$a_2/12 = -a_0/24,\ldots$ which gives $y_1(x) = a_0(1 - x^2/2 -$

$x^3/6 - x^4/24 + \ldots)$.

10. The D.E. transforms to $u''(t) + t^2 u'(t) + (t^2 + 2t)u(t) =$

0. Assuming that $u(t) = \sum\limits_{n=0}^{\infty} a_n t^n$, we have $u'(t) =$

$\sum\limits_{n=1}^{\infty} n a_n t^{n-1}$ and $u''(t) = \sum\limits_{n=2}^{\infty} n(n-1)a_n t^{n-2}$. Substituting in

the D.E. and shifting indices yields

$$\sum_{n=0}^{\infty} (n+2)(n+1)a_{n+2}t^n + \sum_{n=2}^{\infty} (n-1)a_{n-1}t^n + \sum_{n=2}^{\infty} a_{n-2}t^n$$

$$+ \sum_{n=1}^{\infty} 2a_{n-1}t^n = 0,$$

$2 \cdot 1 \cdot a_2 t^0 + (3 \cdot 2 \cdot a_3 + 2 \cdot a_0)t^1 + \sum\limits_{n=2}^{\infty} [(n+2)(n+1)a_{n+2}$

$$+(n+1)a_{n-1} + a_{n-2}]t^n = 0.$$

It follows that $a_2 = 0$, $a_3 = -a_0/3$ and $a_{n+2} = - a_{n-1}/n+2$

$- a_{n-2}/ [(n+2)(n+1)]$, $n = 2,3,4 \ldots$. We obtain one

solution by choosing $a_1 = 0$. Then $a_4 = - a_0/12$, $a_5 =$

$-a_2/5 - a_1/20 = 0$, $a_6 = - a_3/6 - a_2/30 = a_0/18,\ldots$. Thus

one solution is $u_1(t) = a_0(1 - t^3/3 - t^4/12 + t^6/18$

$+ \ldots)$ so $y_1(x) = u_1(x-1) = a_0[1 - (x-1)^3/3 - (x-1)^4/12$

$+ (x-1)^6/18 + \ldots]$. We obtain a second solution by

choosing $a_0 = 0$. Then $a_4 = - a_1/4$, $a_5 = -a_2/5 - a_1/20 =$

$-a_1/20$, $a_6 = -a_3/6 - a_2/30 = 0$, $a_7 = -a_4/7 - a_3/42 =$

$a_1/28,\ldots$. Thus a second linearly independent solution

is $u_2(t) = a_1[t - t^4/4 - t^5/20 + t^7/28 + \ldots]$ or $y_2(x) =$

$u_2(x-1) = a_1[(x-1) - (x-1)^4/4 - (x-1)^5/20$

$+ (x-1)^7/28 + \ldots]$.

The Taylor series for $x^2 - 1$ about $x = 1$ may be obtained

by writing $x = (x-1) + 1$ so $x^2 = (x-1)^2 + 2(x-1) + 1$ and

$x^2 - 1 = (x-1)^2 + 2(x-1)$. The D.E. now appears as $y'' +$

$(x-1)^2 y' + [(x-1)^2 + 2(x-1)]y = 0$ which is identical to

the transformed equation with $t = x - 1$.

12. $y = a_0 + a_1 x + a_2 x^2 + \ldots$, $y^2 = a_0^2 + 2a_0 a_1 x + (2a_0 a_2 +$

$a_1^2)x^2 + \ldots$, $y' = a_1 + 2a_2 x + 3a_3 x^2 + \ldots$, and $(y')^2 =$

$a_1^2 + 4a_1 a_2 x + (6a_1 a_3 + 4a_2^2)x^2 + \ldots$. Substituting

these series in the D.E. and collecting coefficients of

like powers of x yields $(a_1^2 + a_0^2 - 1) + (4a_1 a_2 +$

$2a_0 a_1)x + (6a_1 a_3 + 4a_2^2 + 2a_0 a_2 + a_1^2)x^2 + \ldots = 0$. As in

the earlier problems, each coefficient must be zero. The
I.C. $y(0) = 0$ requires that $a_0 = 0$, and thus $a_1^2 + a_0^2 - 1 = 0$ gives $a_1^2 = 1$. However $y'(0) = a_1 > 0$ implies $a_1 = 1$. Then $4a_1a_2 + 2a_0a_1 = 0$ implies that $a_2 = 0$; and $6a_1a_3 + 4a_2^2 + 2a_0a_2 + a_1^2 = 6a_1a_3 + a_1^2 = 0$ implies that $a_3 = -1/6$. Thus $y = x - x^3/3! + \ldots$.

Section 4.2.1, Page 212

1a. The D.E. can be solved for y'' to yield $y'' = -xy' - y$. If
     $y = \phi(x)$ is a solution, then $\phi''(x) = -x\phi'(x) - \phi(x)$ and
     thus setting $x = 0$ we obtain $\phi''(0) = -0 - 1 = -1$.
     Differentiating the equation for y'' yields $y''' = -xy'' - 2y'$ and hence setting $y = \phi(x)$ again yields $\phi'''(0) = -0 - 0 = 0$. In a similar fashion $y^{iv} = -xy''' - 3y''$ and
     thus $\phi^{iv}(0) = -0 - 3(-1) = 3$. The process can be
     continued to calculate higher derivatives of $\phi(x)$.

2b. The zeros of $P(x) = x^2 - 2x - 3$ are $x = -1$ and $x = 3$.
     For $x_0 = 4$, $x_0 = -4$, and $x_0 = 0$ the distance to the
     nearest zero of $P(x)$ is 1,3, and 1, respectively. Thus a
     lower bound for the radius of convergence for series
     solutions in powers of $(x-4)$, $(x+4)$, and x is $\rho = 1$, $\rho = 3$, and $\rho = 1$, respectively.

4a. If we assume that $y = \sum\limits_{n=0}^{\infty} a_n x^n$, then $y' = \sum\limits_{n=1}^{\infty} na_n x^{n-1}$ and
     $y'' = \sum\limits_{n=2}^{\infty} n(n-1)a_n x^{n-2}$. Substituting in the D.E., shifting

indices of summation, and collecting coefficients of like

powers of x yields the equation

$$(2 \cdot 1 \cdot a_2 + \alpha^2 a_0)x^0 + [3 \cdot 2 \cdot a_3 + (\alpha^2 - 1) a_1]x^1$$

$$+ \sum_{n=2}^{\infty} [(n+2)(n+1)a_{n+2} + (\alpha^2 - n^2)a_n]x^n = 0.$$

Hence the recurrence relation is $a_{n+2} =$

$(n^2 - \alpha^2)a_n/(n+2)(n+1)$, $n = 0, 1, 2, \ldots$. For the first

solution we choose $a_1 = 0$. We find that $a_2 = - \alpha^2 a_0/2 \cdot 1$,

$a_3 = 0$, $a_4 = (2^2 - \alpha^2)a_2/4 \cdot 3 = - (2^2 - \alpha^2)\alpha^2 a_0/4!$ $\ldots$, $a_{2m} =$

$- [(2m-2)^2 - \alpha^2] \ldots (2^2 - \alpha^2)\alpha^2 a_0/(2m)!$, $a_{2m+1} = 0, \ldots$ so

$$y_1(x) = 1 - \frac{\alpha^2}{2!} x^2 - \frac{(2^2 - \alpha^2) \alpha^2}{4!} x^4 - \ldots$$

$$- \frac{[(2m-2)^2 - \alpha^2] \ldots (2^2 - \alpha^2)\alpha^2}{(2m)!} x^{2m} - \ldots$$

where we have set $a_0 = 1$. For the second solution we

take $a_0 = 0$ and $a_1 = 1$ in the recurrence relation to

obtain the desired solution.

4b. If α is an even integer 2k then $(2m-2)^2 - \alpha^2 = (2m-2)^2 -$

4k^2 = 0 when m = k+1. Thus all terms in the series for

$y_1(x)$ are zero after the x^{2k} term. A similar argument

shows that if $\alpha = 2k+1$ then all terms in $y_2(x)$ are zero

after the x^{2k+1} term.

4c. The polynomial solutions for $\alpha = 0, 1, 2, 3, \ldots$ are $y_1(x) =$

1, $y_2(x) = x$, $y_1(x) = 1 - 2x^2$, and $y_2(x) = x - 4x^3/3$,

respectively.

5.   The Taylor series about x = 0 for sin x is sin x = x -

$x^3/3! + x^5/5! - \ldots$. Assuming that $y = \sum\limits_{n=0}^{\infty} a_n x^n$ we find

that

$$y'' + (\sin x)y = 2a_2 + 6a_3 x + 12a_4 x^2 + 20a_5 x^3$$

$$+ 30a_6 x^4 + \ldots + (x-x^3/3! + \ldots)(a_0+a_1 x+a_2 x^2+a_3 x^3+\ldots)$$

$$= 2a_2 + (6a_3+a_0)x + (12a_4+a_1)x^2 + (20a_5+a_2-a_0/6)x^3$$

$$+ (30a_6+a_3-a_1/6)x^4 + \ldots = 0.$$

Hence $a_2 = 0$, $a_3 = - a_0/6$, $a_4 = - a_1/12$, $a_5 = a_0/120$, a_6

$= (a_1+a_0)/180,\ldots$. We set $a_0 = 1$ and $a_1 = 0$ and obtain

$y_1(x) = (1 - x^3/6 + x^5/120 + x^6/180 + \ldots)$. Next we set

$a_0 = 0$ and $a_1 = 1$ and obtain $y_2(x) = (x - x^4/12 + $

$x^6/180 + \ldots)$.

8.   Using the definition of the derivative, we find that

f'(0) and f"(0) are given by:

$$f'(0) = \lim_{h \to 0} \frac{f(h) - f(0)}{h} = \lim_{h \to 0} \frac{\exp(- 1/h^2) - 0}{h}$$

$$= \lim_{h \to 0} \frac{1}{h \exp(1/h^2)} = 0, \text{ using L'Hopital's Rule,}$$

$$f''(0) = \lim_{h \to 0} \frac{f'(h) - f'(0)}{h} = \lim_{h \to 0} \frac{2h^{-3} \exp(-1/h^2) - 0}{h}$$

$$= \lim_{h \to 0} \frac{2}{h^4 \exp(1/h^2)} = 0.$$

9e.   Substituting $y = \sum\limits_{n=0}^{\infty} a_n x^n$ into the D.E. we obtain

$\sum\limits_{n=1}^{\infty} na_n x^{n-1} - \sum\limits_{n=0}^{\infty} a_n x^n = x^2$. Shifting indices in the

first summation and then combining with the second

summation yields $\sum\limits_{n=0}^{\infty} [(n+1) a_{n+1} - a_n]x^n = x^2$. Equating coefficients of both sides then gives: $a_1 - a_0 = 0$, $2a_2 -$

$a_1 = 0$, $3a_3 - a_2 = 1$ and $(n+1)a_{n+1} = a_n$ for $n = 3,4,\ldots$.

Thus $a_1 = a_0$, $a_2 = a_1/2 = a_0/2$, $a_3 = 1/3 + a_2/3 = 1/3 +$

$a_0/2 \cdot 3$, $a_4 = a_3/4 = 1/3 \cdot 4 + a_0/2 \cdot 3 \cdot 4,\ldots$. $a_n = a_{n-1}/n =$

$2/n! + a_0/n!$ and hence $y(x) = a_0(1 + x + \dfrac{x^2}{2!} + \ldots +$

$+ \dfrac{x^n}{n!} \ldots) + 2 \left(\dfrac{x^3}{3!} + \dfrac{x^4}{4!} + \ldots + \dfrac{x^n}{n!} + \ldots\right)$. Using the power

series for e^x, the first and second sums can be rewritten

as desired.

10. Substituting $y = \sum\limits_{n=0}^{\infty} a_n x^n$ into the Legendre equation,

shifting indices, and collecting coefficients of like

powers of x yields

$[2 \cdot 1 \cdot a_2 + \alpha(\alpha+1)a_0]x^0 + \{3 \cdot 2 \cdot a_3 - [2 \cdot 1 - \alpha(\alpha+1)]a_1\}x^1$

$+ \sum\limits_{n=2}^{\infty} \{(n+2)(n+1)a_{n+2} - [n(n+1) - \alpha(\alpha+1)]a_n\}x^n = 0.$

Thus $a_2 = - \alpha(\alpha+1)a_0/2!$, $a_3 = [2 \cdot 1 - \alpha(\alpha+1)]a_1/3! =$

$- (\alpha-1)(\alpha+2)a_1/3!$ and the recurrence relation is

$(n+2)(n+1)a_{n+2} = - [\alpha(\alpha+1) - n(n+1)]a_n = - (\alpha-n) \cdot$

$(\alpha+n+1)a_n$, $n = 2,3,\ldots$. Setting $a_1 = 0$, $a_0 = 1$ yields

a solution with $a_3 = a_5 = a_7 = \ldots = 0$ and $a_4 =$

$\alpha(\alpha-2)(\alpha+1)(\alpha+3)/4!,\ldots$, $a_{2m} = (-1)^m\alpha(\alpha-2)(\alpha-4) \ldots$

$(\alpha-2m+2)(\alpha+1)(\alpha+3) \ldots (\alpha+2m-1)/(2m)!,\ldots$. The second

(linearly independent) solution is obtained by setting a_0

$= 0$ and $a_1 = 1$. The coefficients are $a_2 = a_4 = a_6 = \ldots =$

0 and $a_3 = - (\alpha-1)(\alpha+2)/3!$, $a_5 = - (\alpha-3)(\alpha+4)a_3/5 \cdot 4 =$

$(\alpha-1)(\alpha-3)(\alpha+2)(\alpha+4)/5!,\ldots$.

13. Using the chain rule we have:

$$\frac{dF(\phi)}{d\phi} = \frac{dF[\phi(x)]}{dx}\frac{dx}{d\phi} = -f'(x)\sin\phi(x) = -f'(x)\sqrt{1-x^2} ,$$

$$\frac{d^2F(\phi)}{d\phi^2} = \frac{d}{dx}[-f'(x)\sqrt{1-x^2}]\frac{dx}{d\phi} = (1-x^2)f''(x) - xf'(x),$$

which when substituted into the D.E. yields the desired

result.

15. Carrying out the steps indicated yields the two

equations:

$$P_m[(1-x^2)P_n']' = -n(n+1)P_nP_m$$

$$P_n[(1-x^2)P_m']' = -m(m+1)P_nP_m$$

As long as $n \neq m$ the second equation can be subtracted

from the first and the result integrated from -1 to 1 to

obtain

$$\int_{-1}^{1}\{P_m[(1-x^2)P_n']'-P_n[(1-x^2)P_m']'\}dx = [m(m+1)-n(n+1)]\int_{-1}^{1}P_nP_m dx$$

The left side may be integrated by parts to yield

$$[P_m(1-x^2)P_n' - P_n(1-x^2)P_m']_{-1}^{1} + \int_{-1}^{1}[P_m'(1-x^2)P_n' - $$

$P_n'(1-x^2)P_m']dx$, which is zero. Thus $\int_{-1}^{1}P_n(x)P_m(x)dx = 0$

for $n \neq m$.

Section 4.3, Page 219

1. Since the coefficients of y, y' and y'' have no common

factors and since $P(x)$ vanishes only at $x = 0$ we conclude

that $x = -1$ and $x = 1$ are ordinary points and $x = 0$ is a singular point. Writing the D.E. in the form $y'' + p(x)y'$

$+ q(x)y = 0$, we obtain $p(x) = (1-x)/x$ and $q(x) = 1$. Thus

for the singular point we have

$$\lim_{x \to 0} x\, p(x) = \lim_{x \to 0} 1-x = 1, \lim_{x \to 0} x^2 q(x) = 0.$$

and thus $x = 0$ is a regular singular point.

5. Writing the D.E. in the form $y'' + p(x)y' + q(x)y = 0$, we

find $p(x) = x/(1-x)(1+x)^2$ and $q(x) = 1/(1-x^2)(1+x)$.

Therefore $x = 0$ is an ordinary point and $x = \pm 1$ are

singular points. Since $\lim_{x \to 1} (x-1)p(x)$ and $\lim_{x \to 1} (x-1)^2 q(x)$

both exist, we conclude $x = 1$ is a regular singular

point. Finally, since $\lim_{x \to -1} (x+1)p(x)$ does not exist, we

find that $x = -1$ is an irregular singular point.

10. Writing the D.E. in the form $y'' + p(x)y' + q(x)y = 0$, we

see that $p(x) = e^x/x$ and $q(x) = (3 \cos x)/x$. Thus $x = 0$

is a singular point. Since $xp(x) = e^x$ is analytic at $x =$

0 and $x^2 q(x) = 3x \cos x$ is analytic at $x = 0$ the point x

$= 0$ is a regular singular point.

12. Writing the first D.E. in the form $y'' + p(x)y' + q(x)y =$

0, we see that $p(x) = x/\sin x$ and $q(x) = 4/\sin x$. Since

$\lim_{x \to 0} q(x)$ does not exist the point $x = 0$ is a singular

point. Then $xp(x) = x/f(x)$ is analytic at $x = 0$ and

$x^2 q(x) = 4x/f(x)$ is analytic at $x = 0$ so $x = 0$ is a

regular singular point.

13. If $\xi = 1/x$ then

$$\frac{dy}{dx} = \frac{dy}{d\xi}\frac{d\xi}{dx} = -\frac{1}{x^2}\frac{dy}{d\xi} = -\xi^2\frac{dy}{d\xi} \quad ,$$

$$\frac{d^2y}{dx^2} = \frac{d}{d\xi}\left(-\xi^2\frac{dy}{d\xi}\right)\frac{d\xi}{dx} = \left(-2\xi\frac{dy}{d\xi} - \xi^2\frac{d^2y}{d\xi^2}\right)\left(-\frac{1}{x^2}\right)$$

$$= \xi^4\frac{d^2y}{d\xi^2} + 2\xi^3\frac{dy}{d\xi}.$$

Substituting in the D.E. we have

$$P(1/\xi)[\xi^4\frac{d^2y}{d\xi^2} + 2\xi^3\frac{dy}{d\xi}] + Q(1/\xi)[-\xi^2\frac{dy}{d\xi}] + R(1/\xi)y = 0,$$

$$\xi^4 P(1/\xi)\frac{d^2y}{d\xi^2} + [2\xi^3 P(1/\xi) - \xi^2 Q(1/\xi)]\frac{dy}{d\xi} + R(1/\xi)y = 0.$$

The result then follows from the theory of singular

points at $\xi = 0$.

14b. Since $P(x) = x^2$, $Q(x) = x$, and $R(x) = x^2 - \nu^2$, $f(\xi) =$

$[2P(1/\xi)/\xi - Q(1/\xi)/\xi^2]/P(1/\xi) = 2/\xi - 1/\xi = 1/\xi$ and $g(\xi)$

$= R(1/\xi)/\xi^4 P(1/\xi) = (1/\xi^2 - \nu^2)/\xi^2 = 1/\xi^4 - \nu^2/\xi^2$. Thus

the point at infinity is a singular point. Although $\xi f(\xi)$

$= 1$ is analytic at $\xi = 0$, $\xi^2 g(\xi) = 1/\xi^2 - \nu^2$ is not, so

the point at infinity is an irregular singular point.

Section 4.4, Page 225

2. Assume $y = (x+1)^r$ for $x + 1 > 0$. Substitution of y into

the D.E. yields $[r(r-1) + 3r + 3/4](x+1)^r = 0$. Thus $r^2 +$

$2r + 3/4 = 0$, which yields $r = -3/2, -1/2$. The general

solution of the D.E. is then $y = c_1 |x+1|^{-1/2} +$

$c_2|x+1|^{-3/2}$, $x \neq -1$.

4.   If $y = x^r$ then $r(r-1) + 3r + 5 = 0$. So $r^2 + 2r + 5 = 0$

and $r = (-2 \pm \sqrt{4-20})/2 = -1 \pm 2i$. Thus the general

solution of the D.E. is $y = c_1 x^{-1} \cos(2 \ln|x|) + c_2 x^{-1} \cdot$

$\sin(2 \ln|x|)$, $x \neq 0$.

9.   Again let $y = x^r$ to obtain $r(r-1) - 5r + 9 = 0$, or $(r-3)^2$

$= 0$. Thus the roots are $x = 3,3$ and $y = c_1 x^3 + c_2 x^3 \cdot$

$\ln|x|$, $x \neq 0$, is the solution of the D.E.

13.   Assume that $y = v(x)x^{r_1}$. Then $y' = v(x)r_1 x^{r_1-1} + v'(x)x^{r_1}$

and $y'' = v(x)r_1(r_1-1)x^{r_1-2} + 2v'(x)r_1 x^{r_1-1} + v''(x)x^{r_1}$.

Substituting in the D.E. and collecting terms yields
$x^{r_1+2} v'' + (\alpha + 2r_1)x^{r_1+1} v' + [r_1(r_1-1) + \alpha r_1 + \beta]x^{r_1} v =$

0. Now we make use of the fact that r_1 is a double root

of $f(r) = r(r-1) + \alpha r + \beta$. This means that $f(r_1) = 0$ and

$f'(r_1) = 2r_1 - 1 + \alpha = 0$. Hence the D.E. for v reduces

to $x^{r_1+2} v'' + x^{r_1+1} v'$. Since $x > 0$ we may divide by x^{r_1+1}

to obtain $xv'' + v' = 0$. Thus $v(x) = \ln x$ and a second

solution is $y = x^{r_1} \ln x$.

14.   Substituting $y = x^r$, we find that $r(r-1) + \alpha r + 5/2 = 0$

or $r^2 + (\alpha-1)r + 5/2 = 0$. Thus $r_1, r_2 = [- (\alpha-1) \pm$

$\sqrt{(\alpha-1)^2 - 10}]/2$. In order for solutions to approach

zero as $x \to 0$ it is necessary that the real parts of r_1

and r_2 be positive. Suppose that $\alpha > 1$, then $\sqrt{(\alpha-1)^2} - 10$

is either imaginary or real and less than $\alpha - 1$; hence the real parts of r_1 and r_2 will be negative. Suppose

that $\alpha = 1$, then $r_1, r_2 = \pm i\sqrt{10}$ and the solutions are

oscillatory. Suppose that $\alpha < 1$, then $\sqrt{(\alpha-1)^2 - 10}$ is

either imaginary or real and less than $|\alpha - 1| = 1 - \alpha$;

hence the real parts of r_1 and r_2 will be positive. Thus

if $\alpha < 1$ the solutions of the D.E. will approach zero as

$x \to 0$.

17b. The change of variable $x = e^z$ transforms the D.E. into

u" $- 4u' + 4u = z$, which has the solution $u(z) = c_1 e^{2z} +$

$c_2 z e^{2z} + (1/4)z + 1/4$. Hence $y(x) = c_1 x^2 + c_2 x^2 \ln x +$

$(1/4)\ln x + 1/4$.

19. If $x > 0$, then $|x| = x$ and $|x|^{r_1} = x^{r_1}$ so we can choose c_1

$= k_1$. If $x < 0$, then $|x| = -x$ and $|x|^{r_1} = (-x)^{r_1} = (-1)^{r_1}$.

x^{r_1} and we can choose $c_1 = (-1)^{r_1} k_1$, or $k_1 = (-1)^{r_1} c_1$.

In both cases it is obvious that $c_2 = k_2$.

20. If $y = x^r$ then $r(r-1) + (1-i)r + 2 = 0$. Simiplifying, we

have $r^2 - ir + 2 = (r-2i)(r+i) = 0$, so $r = -i$ and $r = 2i$.

If one does not notice that the quadratic can be

factored, then the quadratic formula can be used: $r = (i$

$\pm \sqrt{-1-8})/2 = (i \pm 3i)/2 = -i$ and $2i$. The general

solution of the D.E. is $y = c_1 x^{2i} + c_2 x^{-i}$, $x \neq 0$. Note

that the solutions are complex-valued functions of the

real variable x.

Section 4.5, Page 232

2.   If the D.E. is put in the standard form y" + p(x)y' +

q(x)y = 0, then p(x) = x^{-1} and q(x) = 1 - $1/9x^2$. Thus x

= 0 is a singular point. Since xp(x) → 1 and x^2q(x) →

-1/9 as x → 0 it follows that x = 0 is a regular singular

point. In determining a series solution of the D.E. it is

more convenient to leave the equation in the form given

rather than divide by the x^2, the coefficient of y". If

we substitute y = $\sum\limits_{n=0}^{\infty} a_n x^{n+r}$, we have

$$\sum\limits_{n=0}^{\infty} (n+r)(n+r-1)a_n x^{n+r} + \sum\limits_{n=0}^{\infty} (n+r)a_n x^{n+r} + (x^2 - \tfrac{1}{9}) \sum\limits_{n=0}^{\infty} a_n x^{n+r} = 0.$$

Note that $x^2 \sum\limits_{n=0}^{\infty} a_n x^{n+r} = \sum\limits_{n=0}^{\infty} a_n x^{n+r+2} = \sum\limits_{n=2}^{\infty} a_{n-2} x^{n+r}$. Thus

we have $[r(r-1) + r - \tfrac{1}{9}]a_0 x^r + [(r+1)r + (r+1) - \tfrac{1}{9}]a_1 x^{r+1}$

$+ \sum\limits_{n=2}^{\infty} \{[(n+r)(n+r-1) + (n+r) - \tfrac{1}{9}]a_n + a_{n-2}\} x^{n+r} = 0.$

The indicial equation is $r^2 - 1/9 = 0$ with roots $r_1 = 1/3$

and $r_2 = -1/3$. For either value of r it is necessary to

take $a_1 = 0$ in order that the coefficient of x^{r+1} be

zero. The recurrence relation is $[(n+r)^2 - 1/9]a_n =$

$-a_{n-2}$. For r = 1/3 we have

$$a_n = \frac{-a_{n-2}}{(n + \tfrac{1}{3})^2 - (\tfrac{1}{3})^2} = -\frac{a_{n-2}}{(n + \tfrac{2}{3})n}, \quad n = 2,3,4,\dots .$$

since $a_1 = 0$ it follows from the recurrence relation that

$a_3 = a_5 = a_7 = \dots = 0$. For the even coefficients it is convenient to let $n = 2m$, $m = 1, 2, 3, \ldots$. Then $a_{2m} = -a_{m-2}/2^2 m(m + \frac{1}{3})$. The first few coefficients are given by

$$a_2 = \frac{(-1)a_0}{2^2(1 + \frac{1}{3})1} \, , \quad a_4 = \frac{(-1)a_2}{2^2(2 + \frac{1}{3})2} = \frac{a_0}{2^4(1 + \frac{1}{3})(2 + \frac{1}{3})2!}$$

$$a_6 = \frac{(-1)a_4}{2^2(3 + \frac{1}{3})3} = \frac{(-1)a_0}{2^6(1 + \frac{1}{3})(2 + \frac{1}{3})(3 + \frac{1}{3})3!} \, ,$$

and the coefficient of x^{2m} for $m = 1, 2, \ldots$ is

$$a_{2m} = \frac{(-1)^m a_0}{2^{2m}m!(1 + \frac{1}{3})(2 + \frac{1}{3}) \ldots (m + \frac{1}{3})} . \quad \text{Thus one}$$

solution (on setting $a_0 = 1$) is

$$y_1(x) = x^{1/3}[1 + \sum_{m=1}^{\infty} \frac{(-1)^m}{m! \, (1 + \frac{1}{3})(2 + \frac{1}{3}) \ldots (m + \frac{1}{3})} (\frac{x}{2})^{2m}] .$$

Since $r_2 = -1/3 \neq r_1$ and $r_1 - r_2 = 2/3$ is not an integer, we can calculate a second series solution corresponding to $r = -1/3$. The recurrence relation is $n(n-2/3)a_n = -a_{n-2}$, which yields the desired solution following the steps just outlined. Note that $a_1 = 0$, as in the first solution, and thus all the odd coefficients are zero.

4. Putting the D.E. in standard form $y'' + p(x)y' + q(x)y = 0$, we see that $p(x) = 1/x$ and $q(x) = -1/x$. Thus $x = 0$ is a singular point, and since $xp(x) \to 1$ and $x^2q(x) \to 0$, as $x \to 0$, $x = 0$ is a regular singular point.

Substituting $y = \sum_{n=0}^{\infty} a_n x^{n+r}$ in $xy'' + y' - y = 0$ and

shifting indices we obtain

$$\sum_{n=-1}^{\infty} a_{n+1}(r+n+1)(r+n)x^{n+r} + \sum_{n=-1}^{\infty} a_{n+1}(r+n+1)x^{n+r} - \sum_{n=0}^{\infty} a_n x^{n+r} = 0.$$

$$[r(r-1) + r]a_0 x^{-1+r} + \sum_{n=0}^{\infty} [(r+n+1)^2 a_{n+1} - a_n]x^{n+r} = 0.$$

The indicial equation is $r^2 = 0$ so $r = 0$ is a double

root. Thus we will obtain only one series of the form y

$= x^r \sum_{n=0}^{\infty} a_n x^n$. The recurrence relation is $(n+1)^2 a_{n+1} = a_n$,

$n = 0,1,2,\ldots$. the coefficients are $a_1 = a_0$, $a_2 = a_1/2^2$

$= a_0/2^2$, $a_3 = a_2/3^2 = a_0/3^2 \cdot 2^2$, $a_4 = a_3/4^2 = a_0/4^2 \cdot 3^2 \cdot 2^2$,

...and $a_n = a_0/(n!)^2$. Thus one solution (on setting $a_0 =$

1) is $y = \sum_{n=0}^{\infty} x^n/(n!)^2$.

7. If we make the change of variable $z = x-1$ and let $y = u(z)$, then the Legendre equation transforms to $(z^2 + 2z)u''(z) + 2(z+1)u'(z) - \alpha(\alpha+1)u(z) = 0$. Since $x = 1$ is

a regular singular point of the original equation, we

know that $z = 0$ is a regular singular point of the

transformed equation. Substituting $u = \sum_{n=0}^{\infty} a_n z^{n+r}$ in the

transformed equation and shifting indices, we obtain

$$\sum_{n=0}^{\infty} (n+r)(n+r-1)a_n z^{n+r} + 2\sum_{n=-1}^{\infty} (n+r+1)(n+r)a_{n+1} z^{n+r}$$

$$+ 2\sum_{n=0}^{\infty} (n+r)a_n z^{n+r} + 2\sum_{n=-1}^{\infty} (n+r+1)a_{n+1} z^{n+r}$$

$$- \alpha(\alpha+1)\sum_{n=0}^{\infty} a_n z^{n+r} = 0,$$

or

$$[2r(r-1) + 2r]a_0 + \sum_{n=0}^{\infty} \{2(n+r+1)^2 a_{n+1}$$

$$+ [(n+r)(n+r+1) - \alpha(\alpha+1)]a_n\}z^{n+r} = 0.$$

The indicial equation is $2r^2 = 0$ so $r = 0$ is a double

root. Thus there will be only one series solution of the

form $y = \sum_{n=0}^{\infty} a_n z^{n+r}$. The recurrence relation is

$2(n+1)^2 a_{n+1} = [\alpha(\alpha+1) - n(n+1)]a_n$, $n = 0,1,2,\ldots$. We have

$a_1 = [\alpha(\alpha+1)]a_0/2 \cdot 1^2$, $a_2 = [\alpha(\alpha+1)][\alpha(\alpha+1) - 1 \cdot 2]a_0/2^2 \cdot 2^2 \cdot$

1^2, $a_3 = [\alpha(\alpha+1)][\alpha(\alpha+1) - 1 \cdot 2][\alpha(\alpha+1) - 2 \cdot 3]a_0/2^3 \cdot 3^2 \cdot 2^2 \cdot$

$1^2,\ldots$, and $a_n = [\alpha(\alpha+1)][\alpha(\alpha+1) - 1 \cdot 2] \ldots [\alpha(\alpha+1) -$

$(n-1)n]a_0/2^n(n!)^2$. Reverting to the variable x it

follows that one solution of the Legendre equation in

powers of x-1 is $y_1(x) = \sum_{n=0}^{\infty} [\alpha(\alpha+1)][\alpha(\alpha+1) - 1 \cdot 2] \ldots$

$[\alpha(\alpha+1) - (n-1)n](x-1)^n/2^n(n!)^2$ where we have set $a_0 = 1$.

10. Putting the D.E. in standard form $y'' + p(x)y' + q(x)y = $

0, we see that $p(x) = 1/x$ and $q(x) = 1$. Thus $x = 0$ is a

singular point; and since $xp(x) \to 1$ and $x^2q(x) \to 0$ as

$x \to 0$, $x = 0$ is a regular singular point. Substituting

$y = \sum_{n=0}^{\infty} a_n x^{n+r}$ into $x^2y'' + xy' + x^2y = 0$ and shifting

indices appropriately, we obtain

$$\sum_{n=0}^{\infty} (n+r)(n+r-1)a_n x^{n+r} + \sum_{n=0}^{\infty} (n+r)a_n x^{n+r} + \sum_{n=2}^{\infty} a_{n-2}x^{n+r} = 0,$$

$$[r(r-1)+r]a_0 x^r + [(1+r)r+1+r]a_1 x^{r+1}$$

$$+ \sum_{n=2}^{\infty} [(n+r)^2 a_n + a_{n-2}]x^{n+r} = 0. \text{ The indicial equation}$$

is $r^2 = 0$ so $r = 0$ is a double root. It is necessary to take $a_1 = 0$ in order that the coefficient of x^{r+1} be zero. The recurrence relation in $n^2 a_n = -a_{n-2}$, $n = 2,3,\ldots$. Since $a_1 = 0$ it follows that $a_3 = a_5 = a_7 = \ldots = 0$. For the even coefficients we let $n = 2m$, $m = 1,2,\ldots$. Then $a_{2m} = -a_{2m-2}/2^2 m^2$ so $a_2 = -a_0/2^2 \cdot 1^2$, $a_4 = a_0/2^2 \cdot 2^2 \cdot 1^2 \cdot 2^2, \ldots$, and $a_{2m} = (-1)^m a_0/2^{2m}(m!)^2$. Thus one solution of the Bessel equation of order zero is $J_0(x) = 1 + \sum\limits_{m=1}^{\infty} (-1)^m x^{2m}/2^{2m}(m!)^2$ where we have set $a_0 = 1$. Using the ratio test it can be shown that the series converges for all x. Also note that $J_0(x) \to 1$ as $x \to 0$.

11. In order to determine the form of the integral for x near zero we must study the integrand for s small. Using the above series for J_0, we have

$$\frac{1}{s[J_0(s)]^2} = \frac{1}{s[1 - s^2/2 + \ldots]^2} = \frac{1}{s[1 - s^2 + \ldots]} =$$

$\frac{1}{s}[1 + s^2 + \ldots]$ for s small. Thus

$$y_2(x) = J_0(x) \int^x \frac{ds}{s[J_0(s)]^2} = J_0(x) \int^x [\frac{1}{s} + s + \ldots]ds$$

$$= J_0(x)[\ln x + \frac{x^2}{2} + \ldots],$$

and it is clear that $y_2(x)$ will contain a logarithmic term.

12a. Putting the D.E. in the standard form $y" + p(x)y' + q(x)y$

= 0 we see that $p(x) = 1/x$ and $q(x) = (x^2-1)/x^2$. Thus x

= 0 is a singular point and since $xp(x) \to 1$ and $x^2q(x) \to$

−1 as $x \to 0$, x = 0 is a regular singular point.

Substituting $y = \sum\limits_{n=0}^{\infty} a_n x^{n+r}$ into $x^2 y" + xy' + (x^2-1)y = 0$,

shifting indices appropriately, and collecting

coefficients of common powers of x we obtain

$[r(r-1) + r - 1]a_0 x^r + [(1+r)r + 1 + r - 1]a_1 x^{r+1}$

$$+ \sum_{n=2}^{\infty} \{[(n+r)^2 - 1]a_n + a_{n-2}\}x^{n+r} = 0.$$

The indicial equation is $r^2-1 = 0$ so the roots are $r_1 = 1$

and $r_2 = -1$. For either value of r it is necessary to

take $a_1 = 0$ in order that the coefficient of x^{r+1} be

zero. The recurrence relation is $[(n+r)^2 - 1]a_n = -a_{n-2}$,

n = 2,3,4... . For r = 1 we have $a_n = -a_{n-2}/[n(n+2)]$, n

= 2,3,4,... . Since $a_1 = 0$ it follows that $a_3 = a_5 = a_7 = $

... = 0. Let n = 2m. Then $a_{2m} = -a_{2m-2}/2^2 m(m+1)$, m =

1,2,..., so $a_2 = -a_0/2^2 \cdot 1 \cdot 2$, $a_4 = -a_2/2^2 \cdot 1 \cdot 2 \cdot 3 = a_0/2^2 \cdot 2^2 \cdot$

$1 \cdot 2 \cdot 2 \cdot 3$,..., and $a_{2m} = (-1)^m a_0/2^{2m} m!(m+1)!$. Thus one

solution (set $a_0 = 1/2$) of the Bessel equation of order

one is $J_1(x) = (x/2) \sum\limits_{n=0}^{\infty} (-1)^n x^{2n}/(n+1)!n!2^{2n}$. The ratio

test shows that the series converges for all x. Also note

that $J_1(x) \to 0$ as $x \to 0$.

12b. For r = −1 the recurrence relation is $[(n-1)^2 - 1]a_n = $

$-a_{n-2}$, $n = 2,3,\ldots$. Substituting $n = 2$ into the relation

yields $[(2-1)^2 - 1]a_2 = 0$ $a_2 = -a_0$. Hence it is

impossible to determine a_2 and consequently impossible to

find a series solution of the form $x^{-1} \sum\limits_{n=0}^{\infty} b_n x^n$.

Section 4.5.1, Page 238

1. The D.E. has the form $P(x)y" + Q(x)y' + R(x)y = 0$ with

    $P(x) = x$, $Q(x) = 2x$, and $R(x) = 6e^x$. From this we find

    $p(x) = Q(x)/P(x) = 2$ and $q(x) = R(x)/P(x) = 6e^x/x$ and

    thus $x = 0$ is a singular point. Since $xp(x) = 2x$ and

    $x^2q(x) = 6xe^x$ are analytic at $x = 0$ we conclude that $x =$

    0 is a regular singular point. Next, we have $xp(x) \to 0 =$

    p_0 and $x^2q(x) \to 0 = q_0$ as $x \to 0$ and thus the indicial

    equation is $r(r-1) + 0 \cdot r + 0 = r^2 - r = 0$, which has the

    roots $r_1 = 1$ and $r_2 = 0$.

3. The equation has the form $P(x)y" + Q(x)y' + R(x)y = 0$

    with $P(x) = x(x-1)$, $Q(x) = 6x^2$ and $R(x) = 3$. Since $P(x)$,

    $Q(x)$, and $R(x)$ are polynomials with no common factors and

    $P(0) = 0$ and $P(1) = 0$, we conclude that $x = 0$ and $x = 1$

    are singular points. The first point, $x = 0$, can be

    shown to be a regular singular point using steps similar

    to those shown in Problem 1. For $x = 1$, we must put the

    D.E. in a form similar to Eq.(1) for this case. To do

    this, divide the D.E. by x and multiply by $(x-1)$ to

    obtain $(x-1)^2 y" + 6x(x-1)y' + \frac{3}{x}(x-1)y = 0$. Comparing this to Eq.(1) we find that $(x-1)p(x) = 6x$ and $(x-1)^2q(x)$

$= 3(x-1)/x$ which are both analytic at $x = 1$ and hence $x =$

1 is a regular singular point. These last two

expressions approach $p_0 = 6$ and $q_0 = 0$ respectively as

$x \to 1$, and thus the indicial equation is $r(r-1) + 6r + 0$

$= r(r+5) = 0$.

9. The D.E. has the form $P(x)y" + Q(x)y' + R(x)y = 0$ with

$P(x) = x^2$, $Q(x) = \sin x$, and $R(x) = -\cos x$. Since $p(x)$

$= Q(x)/P(x) = x^{-2}\sin x$ and $q(x) = R(x)/P(x) = -x^{-2}\cos x$,

$x = 0$ is a singular point. Note that $xp(x) = (\sin x)/x \to$

$1 = p_0$ as $x \to 0$ and $x^2q(x) = -\cos x \to -1 = q_0$ as $x \to 0$.

In order to assert that $x = 0$ is a regular singular point

we must demonstrate that $xp(x)$ and $x^2q(x)$, with $xp(x) = 1$

at $x = 0$ and $q(x) = -1$ at $x = 0$, have convergent power

series (are analytic) about $x = 0$. We know that $\cos x$ is

analytic so we need only consider $(\sin x)/x$. But $\sin x =$

$\sum\limits_{n=0}^{\infty} x^{2n+1}/(2n+1)!$ for $-\infty < x < \infty$ so $(\sin x)/x =$

$\sum\limits_{n=0}^{\infty} x^{2n}/(2n+1)!$ and hence is analytic. We conclude that

$x = 0$ is a regular singular point. It follows that the

indicial equation is $r(r-1) + r - 1 = r^2 - 1 = 0$ and the

roots are $r_1 = 1$, $r_2 = -1$. To find the first few terms

of the solution corresponding to $r_1 = 1$, assume that $y =$

$$x(a_0 + a_1x + a_2x^2 + \ldots) = a_0x + a_1x^2 + a_2x^3 + \ldots .$$

Substituting this series for y in the D.E. and expanding

sin x and cos x about x = 0 yields

$$x^2(2a_1 + 6a_2x + 12a_3x^2 + 20a_4x^3 + \ldots) + (x - x^3/3! +$$

$$x^5/5! + \ldots)(a_0 + 2a_1x + 3a_2x^2 + 4a_3x^3 + 5a_4x^4 + \ldots) -$$

$$(1 - x^2/2! + x^4/4! - \ldots)(a_0x + a_1x^2 + a_2x^3 + a_3x^4 +$$

$a_4x^5 + \ldots) = 0.$ Collecting terms, $(2a_1 + 2a_1 - a_1)x^2 +$

$(6a_2 + 3a_2 - a_0/6 - a_2 + a_0/2)x^3 + (12a_3 + 4a_3 - 2a_1/6 -$

$a_3 + a_1/2)x^4 + (20a_4 + 5a_4 - 3a_2/6 + a_0/120 -$

$a_4 + a_2/2 - a_0/24)x^5 + \ldots = 0.$ Simplifying, $3a_1x^2 + (8a_2$

$+ a_0/3)x^3 + (15a_3 + a_1/6)x^4 + (24a_4 - a_0/30)x^5 + \ldots = 0.$

Thus, $a_1 = 0$, $a_2 = -a_0/4!$, $a_3 = 0$, $a_4 = a_0/6!, \ldots .$

Hence $y_1(x) = x - x^3/4! + x^5/6! + \ldots$ where we have set

$a_0 = 1.$

10. We first write the D.E. in the standard form as given for

Theorem 4.3 except that we are expanding in powers of

(x-1) rather than powers of x: $(x-1)^2y" + (x-1)[(x-1)/$

$2\ln x]y' + [(x-1)^2/\ln x]y = 0.$ Since $\ln 1 = 0$, x = 1 is a

singular point. To show it is a regular singular point of

this D.E. we must show that (x-1)/ln x is analytic at x

= 1; it will then follow that $(x-1)^2/\ln x =$

(x-1)[(x-1)/ln x] is also analytic at x = 1. If we expand

ln x in a Taylor series about x = 1 we find that ln x =

$(x-1) - \frac{1}{2}(x-1)^2 + \frac{1}{3}(x-1)^3 - \ldots$. Thus $(x-1)/\ln x = [1 -$

$\frac{1}{2}(x-1) + \frac{1}{3}(x-1)^2 - \ldots]^{-1} = 1 + \frac{1}{2}(x-1) + \ldots$ has a

power series expansion about $x = 1$, and hence is

analytic. We can use the above result to obtain the

indicial equation at $x = 1$. We have

$(x-1)^2 y'' + (x-1)[\frac{1}{2} + \frac{1}{4}(x-1) + \ldots]y' + [(x-1) + \frac{1}{2}(x-1)^2 +$

$\ldots]y = 0$. Thus $p_0 = 1/2$, $q_0 = 0$ and the indicial

equation is $r(r-1) + r/2 = 0$. Hence $r = 1/2$ and $r = 0$.

In order to find the first three non-zero terms in a

series solution corresponding to $r = 1/2$, it is better to

keep the differential equation in its original form and

to substitute the above power series for $\ln x$:

$[(x-1) - \frac{1}{2}(x-1)^2 + \frac{1}{3}(x-1)^3 - \frac{1}{4}(x-1)^4 + \ldots]y'' + \frac{1}{2}y' + y = 0.$

Next we substitute $y = a_0(x-1)^{1/2} + a_1(x-1)^{3/2} +$

$a_2(x-1)^{5/2} + \ldots$ and collect coefficients of like powers

of $(x-1)$ which are then set equal to zero. This requires

some algebra before we find that $6a_1/4 + 9a_0/8 = 0$ and

$5a_2 + 5a_1/8 - a_0/12 = 0$. These equations yield $a_1 =$

$- 3a_0/4$ and $a_2 = 53a_0/480$. With $a_0 = 1$ we obtain the

solution

$y_1(x) = (x-1)^{1/2} - \frac{3}{4}(x-1)^{3/2} + \frac{53}{480}(x-1)^{5/2} + \ldots$. Since

the radius of convergence of the series for $\ln x$ is 1, we

would expect $\rho = 1$.

12a. If we write the D.E. in the standard form as given in
     Theorem 4.3 we obtain $x^2y'' + x[\alpha/x]y' + [\beta/x]y = 0$ where
     $xp(x) = \alpha/x$ and $x^2q(x) = \beta/x$. Neither of these terms are
     analytic at $x = 0$ so $x = 0$ is an irregular singular
     point.

12b. Substituting $y = x^r \sum\limits_{n=0}^{\infty} a_n x^n$ in $x^3y'' + \alpha xy' + \beta y = 0$ gives

     $$\sum_{n=0}^{\infty} (n+r)(n+r-1)a_n x^{n+r+1} + \alpha \sum_{n=0}^{\infty} (n+r)a_n x^{n+r} + \beta \sum_{n=0}^{\infty} a_n x^{n+r}.$$

     Shifting the index in the first series and collecting

     coefficients of common powers of x we obtain $(\alpha r + \beta)a_0 x^r$

     $+ \sum\limits_{n=1}^{\infty} (n+r-1)(n+r-2)a_{n-1} + [\alpha(n+r) + \beta]a_n \; x^{n+r} = 0.$

     Thus the indicial equation is $\alpha r + \beta = 0$ with the single

     root $r = -\beta/\alpha$.

13a. If we write the D.E. in the standard form as given in
     Theorem 4.3, we obtain $x^2y'' + x[\alpha x^{1-s}]y' + [\beta x^{2-t}]y = 0$.
     Hence $xp(x) = \alpha x^{1-s}$ and $x^2q(x) = \beta x^{2-t}$. If $s > 1$, then
     $xp(x)$ has no limit as $x \to 0$, and if $t > 2$ then $x^2q(x)$ has
     no limit as $x \to 0$. Thus if $s > 1$ or $t > 2$ the point $x = 0$
     is an irregular singular point.

13b. Substituting $y = \sum\limits_{n=0}^{\infty} a_n x^{n+r}$ in the D.E. in standard form
     (above) gives

     $$\sum_{n=0}^{\infty} (n+r)(n+r-1)a_n x^{n+r} + \alpha \sum_{n=0}^{\infty} (n+r)a_n x^{n+r+1-s}$$

     $$+ \beta \sum_{n=0}^{\infty} a_n x^{n+r+2-t} = 0.$$

     If $s = 2$ and $t = 2$ the first term in each of the three

series is $r(r-1)a_0x^r$, $\alpha r a_0 x^{r-1}$, and $\beta a_0 x^r$, respectively.

Thus we must have $\alpha r a_0 = 0$ which requires $r = 0$. Hence

there is at most one solution of the assumed form.

13c. If $s = 1$ and $t = 3$ the first term in each of the three

series is $r(r-1)a_0x^r$, $\alpha r a_0 x^r$, and $\beta a_0 x^{r-1}$, respectively.

Thus we must have $\beta a_0 = 0$. But $\beta \neq 0$ so we take $a_0 = 0$;

it then follows from the recurrence relation that $a_1 = a_2$

$= a_3 = \ldots = 0$, and there is no solution of the assumed

form.

13d. In order for the indicial equation to be quadratic in r

it is necessary that the first term in the first series

contribute to the indicial equation. This means that the

first term in the second and the third series cannot

appear before the first term of the first series. The

first terms are $r(r-1)a_0x^r$, $\alpha r a_0 x^{r+1-s}$, and $\beta a_0 x^{r+2-t}$,

respectively. Thus if $s \leq 1$ and $t \leq 2$ the quadratic term

will appear in the indicial equation.

Section 4.7, Page 253

1a. It is clear that $x = 0$ is a singular point. The D.E. is

in the standard form given in Theorem 4.4 with $xp(x)$

$= 2$ and $x^2q(x) = x$. Both are analytic at $x = 0$, so $x = 0$

is a regular singular point. Substituting $y = \sum\limits_{n=0}^{\infty} a_n x^{n+r}$

in the D.E., shifting indices appropriately, and

collecting coefficients of like powers of x yields

$$[r(r-1) + 2r]a_0 x^r + \sum_{n=1}^{\infty} [(r+n)(r+n+1)a_n + a_{n-1}]x^{r+n} = 0.$$

The indicial equation is $F(r) = r(r+1) = 0$ with roots $r_1 = 0$, $r_2 = -1$. Treating a_n as a function of r, we see that $a_n(r) = -a_{n-1}(r)/F(r+n)$, $n = 1,2,\ldots$ if $F(r+n) \neq 0$. Thus $a_1(r) = -a_0/F(r+1)$, $a_2(r) = a_0/F(r+1)F(r+2),\ldots$, and $a_n(r) = (-1)^n a_0/F(r+1)F(r+2)\ldots F(r+n)$, provided $F(r+n) \neq 0$ for $n = 1,2,\ldots$. For the case $r_1 = 0$, we have $a_n(0) = (-1)^n a_0/F(1)F(2) \ldots F(n) = (-1)^n a_0/n!(n+1)!$ so one solution is $y_1(x) = \sum_{n=0}^{\infty} (-1)^n x^n/n!(n+1)!$ where we have set $a_0 = 1$.

If we try to use the above recurrence relation for the case $r_2 = -1$ we find that $a_n(-1) = -a_{n-1}/n(n-1)$, which is undefined for $n = 1$. Thus we must follow the procedure described at the end of Section 4.6 to calculate a second solution of the form given in Eq.(6b). Specifically, we use Eqs.(15) and (16) of that section to calculate a and $c_n(r_2)$ where $r_2 = -1$. Since $r_1 - r_2 = 1 = N$, we have $a_N(r) = a_1(r) = -1/F(r+1)$. Hence $a = \lim_{r \to -1} [(r+1)(-1)/F(r+1)] = \lim_{r \to -1} [-(r+1)/(r+1)(r+2)] = -1$. Next

$$c_n(-1) = \frac{d}{dr}[(r+1)a_n(r)]\Big|_{r=-1} = (-1)^n \frac{d}{dr}\left[\frac{(r+1)}{F(r+1) \ldots F(r+n)}\right]\Big|_{r=-1},$$

where we have set $a_0 = 1$. Observe that $(r+1)/F(r+1)\ldots F(r+n) = 1/[(r+2)^2(r+3)^2\ldots(r+n)^2(r+n+1)] = 1/G_n(r)$.

Hence $c_n(-1) = (-1)^{n+1} G_n'(-1)/G_n^2(-1)$. Notice that

$G_n(-1) = 1^2 \cdot 2^2 \cdot 3^2 \ldots (n-1)^2$ $n = (n-1)!n!$ and

$G_n'(-1)/G_n(-1) = 2[1/1 + 1/2 + 1/3 + \ldots + 1/(n-1)] + 1/n =$

$H_n + H_{n-1}$. Thus $c_n(-1) = (-1)^{n+1}(H_n + H_{n-1})/(n-1)!n!$

From Eq.(6b) of Section 4.6 we obtain the second solution

$$y_2(x) = - y_1(x) \ln x + x^{-1}[1 - \sum_{n=1}^{\infty} (-1)^n (H_n + H_{n-1})x^n/n!(n-1)!].$$

1b.   It is clear that $x = 0$ is a singular point. The D.E. is

in the standard form given in Theorem 4.4 with $xp(x) = 3$

and $x^2 q(x) = 1+x$. Both are analytic at $x = 0$, so $x = 0$

is a regular singular point. Substituting $y = \sum_{n=0}^{\infty} a_n x^{n+r}$

in the D.E., shifting indices appropriately, and

collecting coefficients of like powers of x yields

$$[r(r-1) + 3r + 1]a_0 x^r + \sum_{n=1}^{\infty} \{[(r+n)(r+n+2) + 1]a_n + a_{n-1}\} x^{n+r} = 0.$$

The indicial equation is $F(r) = r^2 + 2r + 1 = (r+1)^2 = 0$

with the double root $r_1 = r_2 = -1$. Treating a_n as a

function of r, we see that $a_n(r) = -a_{n-1}(r)/F(r+n)$, $n =$

$1,2,\ldots$. Thus $a_1(r) = -a_0/F(r+1)$, $a_2(r) =$

$a_0/F(r+1)F(r+2),\ldots$, and $a_n(r) = (-1)^n a_0/F(r+1)F(r+2)\ldots$

$F(r+n)$. Setting $r = -1$ we find that $a_n(-1) =$

$(-1)^n a_0/(n!)^2$, $n = 1,2,\ldots$. Hence one solution is $y_1(x)$

$= x^{-1} \sum_{n=0}^{\infty} (-1)^n x^n/(n!)^2$ where we have set $a_0 = 1$.

To find a second solution we follow the procedure

described in Section 4.6 for the case when the roots of

the indicial equation are equal. Specifically, the second

solution will have the form given in Eq.(14) of that

section. We must calculate $a_n'(-1)$. If we let $G_n(r) =$

$F(r+1)\ldots F(r+n) = (r+2)^2(r+3)^2\ldots(r+n+1)^2$ and take $a_0 =$

1, then $a_n'(-1) = (-1)^n[1/G_n(r)]'$ evaluated at $r = -1$.

Hence $a_n'(-1) = (-1)^{n+1}G_n'(-1)/G_n^2(-1)$. But $G_n(-1) = (n!)^2$

and $G_n'(-1)/G_n(-1) = 2[1/1 + 1/2 + 1/3 + \ldots +1/n] = 2H_n$.

Thus a second solution is $y_2(x) = y_1(x) \ln x -$

$2x^{-1} \sum_{n=1}^{\infty} (-1)^n H_n x^n/(n!)^2$.

1c. The roots of the indicial equation are r_1 and $r_2 = 0$ and

thus the analysis is similar to that for Problem 1b.

1d. The roots of the indicial equation are $r_1 = -1$ and $r_2 =$

-2 and thus the analysis is similar to that for Problem

1a.

2. Since $x = 0$ is a regular singular point, substitute $y =$

$\sum_{n=0}^{\infty} a_n x^{n+r}$ in the D.E., shift indices appropriately, and

collect coefficients of like powers of x to obtain

$[r^2 - 9/4]a_0 x^r + [(r+1)^2 - 9/4]a_1 x^{r+1}$

$+ \sum_{n=2}^{\infty} \{[(r+n)^2 - 9/4]a_n + a_{n-2}\} x^{n+r} = 0.$

The indicial equation is $F(r) = r^2 - 9/4 = 0$ with roots

$r_1 = 3/2$, $r_2 = -3/2$. Treating a_n as a function of r we

see that $a_n(r) = -a_{n-2}(r)/F(r+n)$, $n = 2,3,\ldots$ if $F(r+n) \neq$

0. For the case $r_1 = 3/2$, $F(r_1+1)$ which is the

coefficient of x^{r_1+1} is $\neq 0$ so we must set $a_1 = 0$. It

follows that $a_3 = a_5 =...= 0$. For the even coefficients,

set $n = 2m$ so $a_{2m}(3/2) = - a_{2m-2}(3/2)/F(3/2 + 2m)$, $m =$

1,2... . Thus $a_2(3/2) = - a_0/2^2 \cdot 1(1 + 3/2)$, $a_4(3/2) =$

$a_0/2^4 \cdot 2!(1 + 3/2)(2 + 3/2),...$, and $a_{2m}(3/2) =$

$(-1)^m/2^{2m}m! \cdot (1 + 3/2)...(m + 3/2)$. Hence one solution

is

$$y_1(x) = x^{3/2}[1 + \sum_{m=1}^{\infty} \frac{(-1)^m}{m!(1 + 3/2)(2 + 3/2)\cdots(m + 3/2)}(\frac{x}{2})^{2m}],$$

where we have set $a_0 = 1$. For this problem, the roots r_1

and r_2 of the indicial equation differ by an integer: r_1

$- r_2 = 3/2 - (-3/2) = 3$. Hence we can anticipate that

there may be difficulty in calculating a second solution

corresponding to $r = r_2$. This difficulty will occur in

calculating $a_3(r) = - a_1(r)/F(r+3)$ because when $r = r_2 =$

$-3/2$ we have $F(r_2+3) = F(r_1) = 0$. However, in this

problem we are fortunate because $a_1 = 0$ and it will not

be necessary to use the theory described at the end of

Section 4.6. Notice for $r = r_2 = -3/2$ that the

coefficient of x^{r_2+1} is $[(r_2+1)^2 - 9/4]a_1$, which does not

vanish unless $a_1 = 0$. Thus the recurrence relation for

the odd coefficients yields $a_5 = -a_3/F(7/2)$, $a_7 =$

$-a_5/F(11/2) = a_3/F(11/2)F(7/2)$ and so forth. Substituting these terms into the assumed form we see that a multiple

of $y_1(x)$ has been obtained and thus we may take $a_3 = 0$

without loss of generality. Hence $a_3 = a_5 = a_7 = \ldots = 0$.

The even coefficients are given by $a_{2m}(-3/2) =$

$- a_{2m-2}(-3/2)/F(2m - 3/2)$, $m = 1,2\ldots$. Thus $a_2(-3/2) =$

$- a_0/2^2 \cdot 1 \cdot (1 - 3/2)$, $a_4(-3/2) = a_0/2^4 \cdot 2!(1 - 3/2)(2 -$

$3/2),\ldots$, and $a_{2m}(-3/2) = (-1)^m a_0/2^{2m} m!(1 - 3/2)(2 - 3/2)$

$\ldots (m - 3/2)$. Thus a second solution is

$$y_2(x) = x^{-3/2}[1 + \sum_{m=1}^{\infty} \frac{(-1)^m}{m!(1 - 3/2)(2 - 3/2) \cdots (m - 3/2)} (\frac{x}{2})^{2m}].$$

4.   Apply the ratio test:

$$\lim_{m \to \infty} \left| \frac{(-1)^{m+1} x^{2m+2}/2^{2m+2}[(m+1)!]^2}{(-1)^m x^{2m}/2^{2m}(m!)^2} \right| = |x|^2 \lim_{m \to \infty} \frac{1}{2^2(m+1)^2} = 0$$

for every x.   Thus the series for $J_0(x)$ converges

absolutely for all x.

9.   If $\xi = \alpha x^\beta$, then $dy/dx = \frac{1}{2}x^{-1/2} f + x^{1/2} f'\alpha\beta x^{\beta-1}$

where f' denotes $df/d\xi$. Find d^2y/dx^2 in a similar

fashion and use algebra to show that f satisfies the D.E.

$\xi^2 f'' + \xi f' + [\xi^2 - \nu^2]f = 0$.

10.   To compare $y'' - xy = 0$ with the D.E. of Problem 9, we

must multiply by x^2 to get $x^2 y'' - x^3 y = 0$. Thus $2\beta = 3$,

$\alpha^2\beta^2 = -1$ and $1/4 - \nu^2\beta^2 = 0$. Hence $\beta = 3/2$, $\alpha = 2i/3$ and

$\nu = 1/3$ which yields the desired result.

11.   First we verify that $J_0(\lambda_j x)$ satisfies the D.E.   We know

that $J_0(t)$ is a solution of the Bessel equation of order

zero:

$$t^2 J_0''(t) + t J_0'(t) + t^2 J_0(t) = 0 \text{ or}$$

$$J_0''(t) + t^{-1} J_0'(t) + J_0(t) = 0.$$

Let $t = \lambda_j x$. Then

$$\frac{d}{dx} J_0(\lambda_j x) = \frac{d}{dt} J_0(t) \frac{dt}{dx} = \lambda_j J_0'(t)$$

$$\frac{d^2}{dx^2} J_0(\lambda_j x) = \lambda_j \frac{d}{dt}[J_0'(t)]\frac{dt}{dx} = \lambda_j^2 J_0''(t).$$

Substituting $y = J_0(\lambda_j x)$ in the given D.E. and making use of these results, we have

$$\lambda_j^2 J_0''(t) + (\lambda_j/t)\,\lambda_j J_0'(t) + \lambda_j^2 J_0(t) =$$

$$\lambda_j^2[J_0''(t) + t^{-1} J_0'(t) + J_0(t)] = 0.$$

Thus $y = J_0(\lambda_j x)$ is a solution of the given D.E. For the second part of the problem we follow the hint. First, rewrite the D.E. by multiplying by x to yield $xy'' + y' + \lambda_j^2 xy = 0$, which can be written as $(xy')' = -\lambda_j^2 xy$. Now let $y_i(x) = J_0(\lambda_i x)$ and $y_j(x) = J_0(\lambda_j x)$ and we have, respectively: $(xy_i')' = -\lambda_i^2 xy_i$

$$(xy_j')' = -\lambda_j^2 xy_j.$$

Now multiply the first equation by y_j, the second by y_i, integrate each from 0 to 1, and subtract the second from the first:

$$\int_0^1 [y_j(xy_i')' - y_i(xy_j')']dx = -(\lambda_i^2 - \lambda_j^2)\int_0^1 xy_i y_j\,dx.$$

If we integrate each term on the left side once by
parts and note that $y_i = y_j = 0$ at $x = 0$ and $x = 1$, we
find that the left side of this equation is identically
zero. Hence the right side is identically zero and for
$\lambda_i \neq \lambda_j$ this gives the desired result.

112

CHAPTER 5

Section 5.1, Page 258

1. Since $y = \phi(x)$ is a solution of the I.V.P. in the interval containing the origin, ϕ must satisfy the D.E. at $x = 0$. Substituting $x = 0$, $\phi(0) = y_0$, $\phi'(0) = y_0', \ldots$, $\phi^{(n-1)}(0) = y_0^{(n-1)}$ and solving for $\phi^{(n)}(0)$ we obtain,
$\phi^{(n)}(0) = -p_1(0)y_0^{(n-1)} - \ldots - p_n(0)y_0$. Since $p_1, \ldots, p_n$ are differentiable at $x = 0$, we differentiate the D.E.
$\phi^{(n)}(x) = -[p_1(x)\phi^{(n-1)}(x) + \ldots + p_n(x)\phi(x)]$ using the product rule and then evaluate at $x = 0$ to obtain
$\phi^{(n+1)}(0) = -p_1(0)\phi^{(n)}(0) - p_1'(0)y_0^{(n-1)} - \ldots - p_n(0)y_0'$
$- p_n'(0)y_0$. Substituting for $\phi^{(n)}(0)$ from above we obtain the desired value for $\phi^{(n+1)}(0)$.

2d. Differentiating $y(x)$ as given we obtain $y' = c_1 + 2c_2x + 3c_3x^2$, $y'' = 2c_2 + 6c_3x$ and $y''' = 6c_3$. The last equation gives $c_3 = y'''/6$ which when substituted into the equation for y'' yields $c_2 = (y'' - xy''')/2$. Substituting for c_2 and c_3 into the equation for y' and solving for c_1 results in $c_1 = y' - xy'' + x^2y'''/2$. Finally inserting these values of c_1, c_2 and c_3 into the equation for $y(x)$ and simplifying yields the desired result.

3b. Writing the equation in standard form, we obtain $y''' + (\sin x/x)y'' + (3/x)y = \cos x/x$. The functions $p_1(x) =$

sin x/x, $p_3(x) = 3/x$ and g(x) = cos x/x have

discontinuities at x = 0. Hence Theorem 5.1 guarantees

that a solution exists for x < 0 or for x > 0.

Section 5.2, Page 261

4c. That e^x, e^{-x}, and e^{-2x} are solutions can be verified by

direct substitution. Computing the Wronskian we obtain,

$$W(e^x, e^{-x}, e^{-2x}) = \begin{vmatrix} e^x & e^{-x} & e^{-2x} \\ e^x & -e^{-x} & -2e^{-2x} \\ e^x & e^{-x} & 4e^{-2x} \end{vmatrix} = -6e^{-2x}.$$

5. To show that the given Wronskian is zero, it is helpful

to note that $(\sin^2 x)' = 2\sin x \cos x = \sin 2x$. This

result can be obtained directly since $\sin^2 x =$

$(1 - \cos 2x)/2 = \frac{1}{10}(5) + (-1/2)\cos 2x$ and hence $\sin^2 x$ is

a linear comination of 5 and cos 2x. Thus the functions

are linearly dependent and their Wronskian is zero.

6c. If we let $L[y] = y^{iv} - 5y'' + 4y$ and if we use the result

of Problem 6b, we have $L[e^{rx}] = (r^4 - 5r^2 + 4)e^{rx}$. Thus

e^{rx} will be a solution of the D.E. provided $(r^2-4)(r^2-1)$

= 0. Solving for r, we obtain the four solutions e^x,

e^{-x}, e^{2x} and e^{-2x}. Since $W(e^x, e^{-x}, e^{2x}, e^{-2x}) \neq 0$, the

four functions form a fundamental set of solutions.

8a. Let y = xv; then y' = xv' + v, y'' = xv'' + 2v', and y''' =

xv''' + 3v''. Substituting in the D.E., we obtain $x^3(xv'''$

+ 3v'') - $3x^2(xv''+2v')$ + 6x(xv'+v) - 6xv = 0, which

reduces to $x^4 v''' = 0$. So $v = c_1 + c_2 x + c_3 x^2$ and $y = c_1 x$ $+ c_2 x^2 + c_3 x^3$. The first term, $c_1 x$, is the solution that we already knew; the solutions x^2 and x^3 are new linearly independent solutions.

Section 5.3, Page 268

1b. If $-1 + i\sqrt{3} = Re^{i\theta}$, then $R = [(-1)^2 + (\sqrt{3})^2]^{1/2} = 2$. The angle θ is given by $R \cos \theta = 2 \cos \theta = -1$ and $R \sin \theta = 2 \sin \theta = \sqrt{3}$. Hence $\cos \theta = -1/2$ and $\sin \theta = \sqrt{3}/2$ which has the solution $\theta = 2\pi/3$. The angle θ is only determined up to an additive integer multiple of $\pm 2\pi$.

2b. Writing $(1-i)$ in the form $Re^{i\theta}$, we obtain $(1-i) = \sqrt{2} \, e^{i(-\pi/4 + 2m\pi)}$ where m is any integer. Hence, $(1-i)^{1/2}$ $= [2^{1/2} \, e^{i(-\pi/4 + 2m\pi)}]^{1/2} = 2^{1/4} \, e^{i(-\pi/8 + m\pi)}$. We obtain the two square roots by setting $m = 0, 1$. They are $2^{1/4}$. $e^{-i\pi/8}$ and $2^{1/4} \, e^{i3\pi/8}$. Note that any other integer value of m gives one of these two values. Also note that $1 - i$ could be written as $1 - i = \sqrt{2} \, e^{i(7\pi/4 + 2m\pi)}$.

4. We look for solutions of the form $y = e^{rx}$. Substituting in the D.E., we obtain the characteristic equation $r^3 - 3r^2 + 3r - 1 = 0$ which has roots $r = 1, 1, 1$. Since the roots are repeated, the general solution is $y = c_1 e^x + c_2 x e^x + c_3 x^2 e^x$.

7. We look for solutions of the form $y = e^{rx}$. Substituting

in the D.E. we obtain the characteristic equation $r^6 + 1$

= 0. The sixth roots of -1 are obtained by setting m

= 0,1,2,3,4,5 in $(-1)^{1/6} = e^{i(\pi+2m\pi)/6}$. They are $e^{i\pi/6}$

= $(\sqrt{3} + i)/2$, $e^{i\pi/2} = i$, $e^{i5\pi/6} = (-\sqrt{3} + i)/2$, $e^{i7\pi/6}$

= $(-\sqrt{3} - i)/2$, $e^{i3\pi/2} = -i$, and $e^{i11\pi/6} = (\sqrt{3} - i)/2$.

Note that there are three pair of conjugate roots. The

general solution is $y = e^{\sqrt{3}\,x/2}[c_1 \cos (x/2) +$

$c_2 \sin (x/2)] + e^{-\sqrt{3}\,x/2}[c_3 \cos (x/2) + c_4 \sin(x/2)]$

$+ c_5 \cos x + c_6 \sin x.$

15. The characteristic equation is $r^3 + r = 0$ and hence r =

0, $+ i$, $-i$ are the roots and the general solution is y(x)

= $c_1 + c_2 \cos x + c_3 \sin x$. y(0) = 0 implies $c_1 + c_2$ =

0, y'(0) = 1 implies c_3 = 1 and y"(0) = 2 implies $-c_2$ =

2. Use this last equation in the first to find c_1 = 2

and thus y(x) = 2 - 2 cos x + sin x.

17. The general solution would normally be written y(x) = c_1

$+ c_2 x + c_3 e^{2x} + c_4 x e^{2x}$. However, in order to evaluate

the c's when the initial conditions are given at x = 1,

it is advantageous to rewrite y(x) as y(x) = $c_1 + c_2 x$

$+ c_5 e^{2(x-1)} + c_6(x-1)e^{2(x-1)}$.

19. The above approach for solving the D.E. would normally

yield y(x) = $c_1 \cos x + c_2 \sin x + c_5 e^x + c_6 e^{-x}$ as the

solution. Now use the definition of cosh x and sinh x to

yield the desired result.

It is convenient to use cosh x and sinh x rather than e^x and e^{-x} because the I.C. are given at $x = 0$. Since cosh x and sinh x and all of their derivatives are either 0 or 1 at $x = 0$, the algebra in satisfying the I.C. is simplified.

20b. If r_1 is an s-fold root of $F(r)$, $s \geq 2$, then $L[x^r] = x^r(r-r_1)^s H(r)$ where $H(r)$ is a polynomial in r of degree n-s and $H(r_1) \neq 0$. Clearly x^{r_1} is a solution of the D.E. Using the fact that $\partial x^r / \partial r = x^r \ln x$ and that differentiation with respect to x and r may be interchanged, we obtain $\partial L[x^r]/\partial r = L[x^r \ln x] = x^r[\ln x \ (r-r_1)^s \ H(r) + s(r-r_1)^{s-1} \ H(r) + (r-r_1)^s H'(r)]$. Since $s \geq 2$, the right side of this equation vanishes at $r = r_1$, and hence $x^{r_1} \ln x$ is also a solution of the nth order Euler equation. If $s \geq 3$, differentiating $L[x^r \ln x]$ again with respect to r and setting $r = r_1$ shows that $x^{r_1}(\ln x)^2$ is also a solution of the D.E. This process can be continued as long as there is a common power of $(r-r_1)$ on the right side of the D.E.; namely through s-1 differentiations.

21c. We look for solutions of the form $y = x^r$, and obtain $L[x^r] = x^r[r(r-1)(r-2) + 2(r)(r-1) + (r-1)] = 0$. The cubic polynomial in r has a common factor $(r-1)$ and has roots 1, i, and -i. The general solution is $y = c_1 x +$

$c_2 \cos (\ln x) + c_3 \sin (\ln x)$.

23. Let $y = w(z)$ and use the facts $z = \ln x$ and $dy/dx = (dw/dz)(dz/dx)$ to obtain $dy/dx = (x^{-1})w'$, $d^2y/dx^2 = (x^{-2}) \cdot (w''-w')$ and $d^3y/dx^3 = (x^{-3})(w''' - 3w'' + 2w')$ where $'$ stands for d/dz. Substituting in the D.E. gives the third order linear equation with constant coefficients, $w''' - 3w'' + 3w' - 1 = 0$. The characteristic equation $r^3 - 3r^2 + 3r - 1 = 0$ has a triple root, $r = 1$. Hence the general solution is $w = c_1 e^z + c_2 z e^z + c_3 z^2 e^z$. Finally substituting $z = \ln x$, we obtain $y = c_1 x + c_2 x \ln x + c_3 x(\ln x)^2$.

24a. Since $a_0 = 1 > 0$, the quantities that must be checked are

$$3, \quad \begin{vmatrix} 3 & 1 \\ 1 & 3 \end{vmatrix}, \quad \text{and} \quad \begin{vmatrix} 3 & 1 & 0 \\ 1 & 3 & 0 \\ 0 & 0 & 1 \end{vmatrix}.$$

They are each positive; hence the D.E. is asymptotically stable.

24e. Again $a_0 > 0$ and the quantities that must be checked are

$$0.1, \quad \begin{vmatrix} 0.1 & 1 \\ -0.4 & 1.2 \end{vmatrix}, \quad \text{and} \quad \begin{vmatrix} 0.1 & 1 & 0 \\ -0.4 & 1.2 & .1 \\ 0 & 0 & -0.4 \end{vmatrix}.$$

The last quantity has the value $-0.4(0.12 + 0.4) < 0$. Hence the D.E. is unstable.

Section 5.4, Page 273

1. First solve the homogeneous D.E. The characteristic

    equation is $r^3 - r^2 - r + 1 = 0$, and the roots are r =

    -1, 1, 1; hence $y_c(x) = c_1 e^{-x} + c_2 e^x + c_3 x e^x$. Using the

    superposition principle, we can write a particular

    solution as the sum of particular solutions corresponding

    to the D.E. $y''' - y'' - y' + y = 2e^{-x}$ and $y''' - y'' - y' +$

    $y = 3$. Our initial choice for a particular solution,

    y_{p_1}, of the first equation is Ae^{-x}; but e^{-x} is a solution

    of the homogeneous equation so we multiply by x. Thus,

    $y_{p_1}(x) = Axe^{-x}$. For the second equation we choose $y_{p_2}(x)$

    = B, and there is no need to modify this choice. The

    constants are determined by substituting into the

    individual equations. We obtain A = 1/2, B = 3. Thus,

    the general solution is $y = c_1 e^{-x} + c_2 e^x + c_3 x e^x + 3 +$

    $(xe^{-x})/2$.

9. The characteristic equation for the related homogeneous

    D.E. is $r^3 + 4r = 0$ with roots r = 0, +2i, -2i. Hence

    $y_c(x) = c_1 + c_2 \cos 2x + c_3 \sin 2x$. The initial choice

    for $y_p(x)$ is Ax + B, but since B is a solution of the

    homogeneous equation we must multiply by x and assume

    $y_p(x) = x(Ax+B)$. A and B are found by substituting $y_p(x)$

    in the D.E., which gives A = 1/8, B = 0, and thus the

    general solution is $y(x) = c_1 + c_2 \cos 2x + c_3 \sin 2x +$

$(1/8)x^2$. Applying the I.C. we have $y(0) = 0 \to c_1 + c_2 = 0$, $y'(0) = 0 \to 2c_3 = 0$, and $y''(0) = 1 \to -4c_2 + 1/4 = 1$, which have the solution $c_1 = 3/16$, $c_2 = -3/16$, $c_3 = 0$.

12. The characteristic equation for the homogeneous D.E. is $r^3 - 2r^2 + r = 0$ with roots $r = 0,1,1$. Hence the complementary solution is $y_c(x) = c_1 + c_2 e^x + c_3 x e^x$. We consider the differential equations $y''' - 2y'' + y' = x^3$ and $y''' - 2y'' + y' = 2e^x$ separately. Our initial choice for a particular solution, y_{p_1}, of the first equation is $A_0 x^3 + A_1 x^2 + A_2 x + A_3$; but since a constant is a solution of the homogeneous equation we must multiply by x. Thus, $y_{p_1}(x) = x(A_0 x^3 + A_1 x^2 + A_2 x + A_3)$. For the second equation we first choose $y_{p_2}(x) = Be^x$, but since both e^x and xe^x are solutions of the homogeneous equation, we multiply by x^2 to obtain $y_{p_2}(x) = Bx^2 e^x$. Then $y_p(x) = y_{p_1}(x) + y_{p_2}(x)$ by the superposition principle and $y(x) = y_c(x) + y_p(x)$.

16. The complementary solution is $y_c(x) = c_1 + c_2 e^{-x} + c_3 e^x + c_4 x e^x$. The superposition principle allows us to consider separately the D.E. $y^{iv} - y''' - y'' + y' = x^2 + 4$ and $y^{iv} - y''' - y'' + y' = x \sin x$. For the first equation our initial choice is $y_{p_1}(x) = A_0 x^2 + A_1 x + A_2$; but this must be multiplied by x since a constant is a solution of the homogeneous D.E. Hence $y_{p_1}(x) = x(A_0 x^2 + A_1 x + A_2)$. For

the second equation our initial choice that $y_{p_2}(x) =$

$(B_0 x + B_1)\cos x + (C_0 x + C_1)\sin x$ does not need to be

modified. Hence, $y_p(x) = x(A_0 x^2 + A_1 x + A_2) +$

$(B_0 x + B_1)\cos x + (C_0 x + C_1)\sin x$.

19. $(D-a)(D-b)f = (D-a)(Df-bf) = D^2 f - (a+b)Df + abf$ and

$(D-b)(D-a)f = (D-b)(Df-af) = D^2 f - (b+a)Df + baf$. Since

$a+b = b+a$ and $ab = ba$, we find the given equation holds

for any function f.

21a. The D.E. of Problem 12 can be written as $D(D-1)^2 y = x^3 +$

$2e^x$. Since D^4 annihilates x^3 and $(D-1)$ annihilates $2e^x$,

we have $D^5(D-1)^3 y = 0$, which corresponds to Eq.(ii) of

Problem 20. The solution of this equation is $y(x) = A_1 x^4$

$+ A_2 x^3 + A_3 x^2 + A_4 x + A_5 + (B_1 x^2 + B_2 x + B_3)e^{-x}$. Since

$A_5 + (B_2 x + B_3)e^{-x}$ are solutions of the homogeneous

equation related to the original D.E., they may be

deleted and thus $y_p(x) = A_1 x^4 + A_2 x^3 + A_3 x^2 + A_4 x +$

$B_1 x^2 e^{-x}$.

21b. $(D+1)^2(D^2+1)$ annihilates the right side of the D.E. of

Problem 13.

21e. $D^3(D^2+1)^2$ annihilates the right side of the D.E. of

Problem 16.

Section 5.5, Page 278

1.   The complementary solution is $y_c = c_1 + c_2 \cos x +$

c_3 sin x and thus we assume a particular solution of the

form $y_p = u_1(x) + u_2(x) \cos x + u_3(x) \sin x$.

Differentiating y_p and assuming Eq.(5), we obtain $y_p' =$

$-u_2 \sin x + u_3 \cos x$ and

$$u_1' + u_2' \cos x + u_3' \sin x = 0 \qquad (a).$$

Continuing this process we obtain $y_p'' = -u_2 \cos x -$

$u_3 \sin x$, $y_p''' = u_2 \sin x - u_3 \cos x - u_2' \cos x -$

$u_3' \sin x$ and

$$-u_2' \sin x + u_3' \cos x = 0 \qquad (b).$$

Substituting y_p and its derivatives, as given above, into

the D.E. we obtain the third equation:

$$-u_2' \cos x - u_3' \sin x = \tan x \qquad (c).$$

Equations (a),(b) and (c) constitute Eqs.(10) of the text

for this problem and may be solved for $u_1' = \tan x$, $u_2' =$

$-\sin x$, and $u_3' = -\sin^2 x / \cos x$.

5.   Since a fundamental set of solutions of the homogeneous

D.E. is $y_1 = e^x$, $y_2 = \cos x$, $y_3 = \sin x$ a particular

solution is of the form $y_p(x) = e^x u_1(x) + (\cos x)u_2(x) +$

$(\sin x)u_3(x)$. Differentiating and making the same

assumptions that lead to Eqs.(10), we obtain

$$u_1' e^x + u_2' \cos x + u_3' \sin x = 0$$
$$u_1' e^x - u_2' \sin x + u_3' \cos x = 0$$
$$u_1' e^x - u_2' \cos x - u_3' \sin x = g(x).$$

Solving these equations using either determinants or by

elimination, we obtain $u_1' = (1/2)e^{-x}g(x)$, $u_2' =$

$(1/2)(\sin x - \cos x)g(x)$ and $u_3' = -(1/2)(\sin x + \cos x)g(x)$.

Integrating these and substituting into y_p yields

$$y_p(x) = \frac{1}{2} \{ e^x \int^x e^{-t}g(t)dt + \cos x \int^x (\sin t - \cos t)g(t)dt$$
$$- \sin x \int^x (\sin t + \cos t)g(t) \, dt \} .$$

This can be written in the form

$$y_p(x) = (1/2) \int^x (e^{x-t} + \cos x \sin t - \cos x \cos t$$
$$- \sin x \sin t - \sin x \cos t)g(t)dt.$$

If we use the trigonometric identities $\sin(A-B) = \sin A \cdot$

$\cos B - \cos A \sin B$ and $\cos(A-B) = \cos A \cos B + \sin A \cdot$

$\sin B$, we obtain the desired result. Note: Eqs.(11) and

(12) of this section give the same result, but it is not

recommended to memorize these equations.

7. The particular solution has the form $y_p = e^x u_1(x) +$

$xe^x u_2(x) + x^2 e^x u_3(x)$. Differentiating, making the same

assumptions as in the earlier problems, and solving the

three linear equations for u_1', u_2' and u_3' yield $u_1' =$

$(1/2)x^2 e^{-x}g(x)$, $u_2' = -xe^{-x}g(x)$ and $u_3' = (1/2)e^{-x}g(x)$.

Integration and substituting into y_p yields the desired

solution. If $g(x) = x^{-2}e^x$ then $g(t) = e^t/t^2$ and the

integration is accomplished using the power rule.

CHAPTER 6

Section 6.1, Page 284

1a. The graph of $f(t)$ is shown. Since the function is continuous on each interval, but has a jump discontinuity at $t = 1$, $f(t)$ is piecewise continuous.

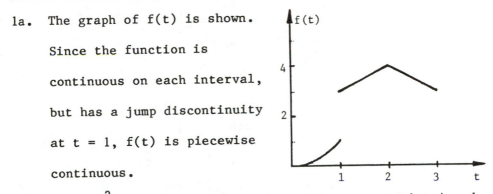

2b. Since t^2 is continuous for $0 \leq t \leq A$ for any postive A and since $t^2 \leq e^{at}$ for any $a > 0$ and for t sufficiently large, it follows from Theorem 6.2 that $\mathcal{L}\{t^2\}$ exists for

$$s > 0. \quad \mathcal{L}\{t^2\} = \int_0^\infty e^{-st}t^2 dt = \lim_{M \to \infty} \int_0^M e^{-st}t^2 dt$$

$$= \lim_{M \to \infty} [\frac{-t^2}{s}e^{-st}\Big|_0^M + \frac{2}{s}\int_0^M e^{-st}t dt]$$

$$= \frac{2}{s} \lim_{M \to \infty} [-\frac{1}{s}te^{-st}\Big|_0^M + \frac{1}{s}\int_0^M e^{-st}dt]$$

$$= \frac{2}{s^2} \lim_{M \to \infty} -\frac{1}{s}e^{-st}\Big|_0^M = \frac{2}{s^3}$$

3. That $f(t) = \cos$ at satisfies the hypotheses of Theorem 6.2 can be verified by recalling that $|\cos at| \leq 1$ for all t. To determine $\mathcal{L}\{\cos at\} = \int_0^\infty e^{-st} \cos at\ dt$ we must integrate by parts twice to get $\int_0^\infty e^{-st} \cos at\ dt =$ $\lim_{M \to \infty} [(-s^{-1} e^{-st} \cos at + as^{-2} e^{-st} \sin at)\Big|_0^M - (a^2/s^2) \cdot$

$\int_0^M e^{-st} \cos at \, dt]$. Evaluating the first two terms,

letting $M \to \infty$, and adding the third term to both sides,

we obtain $[1 + a^2/s^2] \int_0^\infty e^{-st} \cos at \, dt = 1/s$, $s > 0$.

Division by $[1 + a^2/s^2]$ and simplification yields the

desired solution.

4c. From the definition for cosh bt we have

$$\mathcal{L}\{e^{at}\cosh bt\} = \mathcal{L}\{\tfrac{1}{2}[e^{(a+b)t} + e^{(a-b)t}]\} \ .$$

Using the linearity property of $\mathcal{L}$, Eq.(5), the right side

becomes $\tfrac{1}{2}\mathcal{L}\{e^{(a+b)t}\} + \tfrac{1}{2}\mathcal{L}\{e^{(a-b)t}\}$ which can be

evaluated using the result of Example 5 and thus

$$\mathcal{L}\{e^{at}\cosh bt\} = \frac{1/2}{s-(a+b)} + \frac{1/2}{s-(a-b)}$$

$$= \frac{s-a}{(s-a)^2-b^2} \ , \text{ for } s-a > |b|.$$

5c. The linearity of the Laplace transform operator allows us

to write $\mathcal{L}\{e^{at} \sin bt\} = (1/2i) \mathcal{L}\{e^{(a+ib)t}\}$

$- (1/2i)\mathcal{L}\{e^{(a-ib)t}\}$. Each of these two terms can be

evaluated by using the result of Example 5, where we now

have to require s to be greater than the real part of the

complex numbers $a + ib$ and $a - ib$ in order for the

integrals to converge. Complex algebra then gives the

desired result. An alternate method of evaluation would

be to use integration on the integral appearing in the

definition of $\mathcal{L}\{e^{at} \sin bt\}$, but that method is very cumbersome.

6b. As in Problem 5(c) we write sin at = $(e^{iat} - e^{-iat})/2i$

and proceed using the linearity of the Laplace transform

operator. Thus, $\mathcal{L}\{t \sin at\}$ = $(1/2i)\mathcal{L}\{te^{iat}\}$ - $(1/2i)$ ·

$\mathcal{L}\{te^{-iat}\}$. Using the result of part (a) of this problem

we obtain $\mathcal{L}\{t \sin at\}$ = $(1/2i)[(s-b)^{-2} - (s+b)^{-2}]$ where

b = ia and s > 0. Hence $\mathcal{L}\{t \sin at\}$ = $2as/(s^2+a^2)^2, s > 0$.

6e. Use the approach shown in Problem 6(b), with a second

integration by parts.

8.   If we let u = f and dv = $e^{-st}dt$, then F(s) =

$\int_0^\infty e^{-st}f(t)dt = \lim_{M \to \infty} -\frac{1}{5} e^{-st}f(t)\Big|_0^M + \frac{1}{s} \int_0^\infty e^{-st}f'(t)dt.$

Now use an argument similar to that given to establish

Theorem 6.2.

10a. Make a transformation of variables with x = st and dx =

sdt. Then use the definition of $\Gamma(p+1)$ from Problem 9.

10d. Use the definition of $\mathcal{L}\{t^{1/2}\}$ and integrate by parts

once to get $\mathcal{L}\{t^{1/2}\}$ = $(1/2s)\mathcal{L}\{t^{-1/2}\}$. The result

follows from part (c).

Section 6.2, Page 294

Problems 1 through 10 are solved by using partial fractions
and algebra to manipulate the given function into a form
matching one of the functions appearing in the middle column
of Table 6.1.

2.   We have $\dfrac{4}{(s-1)^3}$ = $2\dfrac{2!}{(s-1)^{2+1}}$ and thus the inverse Laplace

transform is $2t^2e^t$.

4. We have $\dfrac{3s}{s^2-s-6} = \dfrac{3s}{(s-3)(s+2)} = \dfrac{9/5}{s-3} + \dfrac{6/5}{s+2}$ using partial

fractions. Thus $(9/5)e^{3t} + (6/5)e^{-2t}$ is the inverse

transform.

7. We have $\dfrac{2s+1}{s^2-2s+2} = \dfrac{2s+1}{(s-1)^2+1} = \dfrac{2(s-1)}{(s-1)^2+1} + \dfrac{3}{(s-1)^2+1}$,

where we first used the concept of completing the square

(in the denominator) and then added and subtracted

appropriately to put the numerator in the desired form.

Table 6.1 may now be used to find the desired result.

In each of the Problems 11 through 23 it is assumed that the
I.V.P. has a solution y = $\phi(t)$ which, with its first two
derivatives, satisfies the conditions of the Corollary to
Theorem 6.3.

11. Take the Laplace transform of the D.E. to get $s^2Y(s)$ −

sy(0) − y'(0) − [sY(s) − y(0)] − 6Y(s) = 0. Using the

I.C. and solving for Y(s) we obtain $Y(s) = \dfrac{s-2}{s^2-s-6}$.

Following the pattern of Eq.(12) we have $\dfrac{s-2}{s^2-s-6} = \dfrac{a}{s+2} +$

$\dfrac{b}{s-3} = \dfrac{a(s-3)+b(s+2)}{(s+2)(s-3)}$. Equating like powers in the

numerators we find a+b = 1 and −3a + 2b = −2. Thus a =

4/5 and b = 1/5 and $Y(s) = \dfrac{4/5}{s+2} + \dfrac{1/5}{s-3}$, which yields the

desired solution using Table 6.1.

14. Taking the Laplace transform we have $s^2Y(s)$ − sy(0)−y'(0)

−4[sY(s)−y(0)] + 4Y(s) = 0. Using the I.C. and solving

for Y(s) we find $Y(s) = \dfrac{s-3}{s^2-4s+4}$. Since the denominator is a perfect square, the partial fraction form is $\dfrac{s-3}{s^2-4s+4}$

$= \dfrac{a}{(s-2)^2} + \dfrac{b}{s-2}$. Solving for a and b, as shown in

examples of this section or in Problem 11, we find a =

-1 and b = 1. Thus $Y(s) = \dfrac{1}{s-2} - \dfrac{1}{(s-2)^2}$, from which

we find $y(t) = e^{2t} - te^{2t}$.

15.   Note that $Y(s) = \dfrac{2s-4}{s^2-2s-2} = \dfrac{2s-4}{(s-1)^2-3} = \dfrac{2(s-1)}{(s-1)^2-3} - \dfrac{2}{(s-1)^2-3}$.

Three formulas in Table 6.1 are now needed; F(s-c) in

conjunction with the ones for cosh at and sinh at.

17.   The Laplace transform of the D.E. is

$s^4Y(s) - s^3y(0) - s^2y'(0) - sy''(0) - y'''(0) - 4[s^3Y(s)-s^2y(0)$

$-sy'(0) - y''(0)] + 6[s^2Y(s) - sy(0) - y'(0)] - 4[sY(s) - y(0)]$

$+ Y(s) = 0$. Using the I.C. and solving for Y(s) we find

$Y(s) = \dfrac{s^2 - 4s + 7}{s^4-4s^3+6s^2-4s+1}$. The correct partial fraction

form for this is $\dfrac{a}{(s-1)^4} + \dfrac{b}{(s-1)^3} + \dfrac{c}{(s-1)^2} + \dfrac{d}{s-1}$.

Setting this equal to Y(s) above and equating the

numerators we have $s^2-4s+7 = a + b(s-1) + c(s-1)^2 +$

$d(s-1)^3$. Solving for a,b,c, and d and use of Table 6.1

yields the desired solution.

20.   The Laplace transform of the D.E. is $s^2Y(s) - sy(0)$

$- y'(0) + \omega^2Y(s) = s/(s^2+4)$. Applying the I.C. and

solving for Y(s) we get $Y(s) = s/[(s^2+4)(s^2+\omega^2)]$

$+ s/(s^2+\omega^2)$. Decomposing the first term by partial

fractions we have $Y(s) = s/[(\omega^2-4)(s^2+4)] - s/[(\omega^2-4)$.

$(s^2+\omega^2] + s/(s^2+\omega^2)$. Combining the last two terms we can write $Y(s) = (\omega^2-4)^{-1}[(\omega^2-5)s/(s^2+\omega^2) + s/(s^2+4)]$. Then, using Table 6.1, we have $y = (\omega^2-4)^{-1}[(\omega^2-5)\cos \omega t + \cos 2t]$.

22. Solving for $Y(s)$ we find $Y(s) = 1/[(s-1)^2 + 1] + 1/(s+1)[(s-1)^2 + 1]$. Using partial fractions on the second term we obtain $Y(s) = 1/[(s-1)^2 + 1] + \{1/(s+1) - (s-3)/[(s-1)^2 + 1]\}/5 = (1/5)\{(s+1)^{-1}-(s-1)[(s-1)^2 + 1]^{-1} + 7[(s-1)^2 + 1]^{-1}\}$. Hence, $y = (1/5)(e^{-t} - e^t\cos t + 7e^t \sin t)$.

24. Under the standard assumptions, the Laplace transform of the left side of the D.E. is $s^2 Y(s) - sy(0) - y'(0) + 4Y(s)$. To transform the right side, we must revert to the definition of the Laplace transform to determine $\int_0^\infty e^{-st}f(t)dt$. Since $f(t)$ is piecewise continuous we are able to calculate $\mathcal{L}\{f(t)\}$ by

$$\int_0^\infty e^{-st} f(t)dt = \int_0^\pi e^{-st} dt + \lim_{M\to\infty} \int_\pi^M (e^{-st})(0)dt =$$

$$\int_0^\pi e^{-st}dt = (1 - e^{-\pi s})/s.$$

Hence, the Laplace transform $Y(s)$ of the solution is given by $Y(s) = s/(s^2+4) + (1 - e^{-\pi s})/s(s^2+4)$.

27b. The Taylor series for f about $t = 0$ is $f(t) = \sum_{n=0}^\infty (-1)^n t^{2n}/(2n+1)!$, which is obtained from part (a) by dividing each term of the sine series by t. Also, f is continuous for $t > 0$ since $\lim_{t \to 0+} (\sin t)/t = 1$. Assuming

that we can compute the Laplace transform of f term by

term, we obtain $\mathcal{L}\{f(t)\} = \mathcal{L}\{ \sum_{n=0}^{\infty} (-1)^n t^{2n}/(2n+1)! \}$

$= \sum_{n=0}^{\infty} [(-1)^n/(2n+1)!] \mathcal{L}\{t^{2n}\}$

$= \sum_{n=0}^{\infty} [(-1)^n(2n)!/(2n+1)!]s^{-(2n+1)}$

$= \sum_{n=0}^{\infty} [(-1)^n/(2n+1)]s^{-(2n+1)}$, which converges for $s > 1$.

The Taylor series for arctan x is given by $\sum_{n=0}^{\infty} (-1)^n \cdot$

$x^{2n+1}/(2n+1)$. Comparing $\mathcal{L}\{f(t)\}$ with the Taylor series

for arctan x, we conclude that $\mathcal{L}\{f(t)\} = \arctan(1/s)$,

$s > 1$.

29b. Setting n = 2 in Problem 28b. we have

$$\mathcal{L}\{t^2 \sin bt\} = \frac{d^2}{ds^2} [b/(s^2+b^2)] = \frac{d}{ds} [-2bs/(s^2+b^2)^2] =$$

$$-2b/(s^2+b^2)^2 + 8bs^2/(s^2+b^2)^3 = 2b(3s^2-b^2)/(s^2+b^2)^3.$$

29d. Using the result of Problem 28a. we have

$$\mathcal{L}\{te^{at}\} = -\frac{d}{ds} (s-a)^{-1} = (s-a)^{-2}$$

$$\mathcal{L}\{t^2 e^{at}\} = -\frac{d}{ds} (s-a)^{-2} = 2(s-a)^{-3}.$$

$$\mathcal{L}\{t^3 e^{at}\} = -\frac{d}{ds} 2(s-a)^{-3} = 3!(s-a)^{-4}.$$ Continuing in

this fashion or using induction we obtain the desired

result.

31a. Taking the Laplace transform of the D.E. we obtain

$$\mathcal{L}\{y''\} + \mathcal{L}\{ty\} = \mathcal{L}\{y''\} - \mathcal{L}\{-ty\} = s^2 Y(s) - sy(0) - y'(0)$$

$- Y'(s) = 0$. Hence, Y satisfies $Y' - s^2 Y = -s$.

33a. Follow the hint and apply L'Hopital's rule recalling that

      $Q(s)$ has distinct zeros.

Section 6.3, Page 302

1b. From the definition of $u_c(t)$ we have

$$g(t) = (t-3)u_2(t) - (t-2)u_3(t) = \begin{cases} 0 - 0 = 0, & 0 \le t < 2 \\ (t-3) - 0 = t-3, & 2 \le t < 3. \\ (t-3) - (t-2) = -1, & 3 \le t \end{cases}$$

    The graph of $y = g(t)$ is:

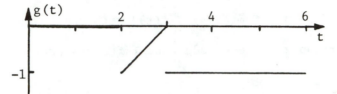

1d. As indicated in the discussion following Example 1, the

    unit step function can be used to translate a given

    function f, with domain $t \ge 0$, a distance c to the right

    by the multiplication $u_c(t)f(t-c)$. Hence the required

    graph of $y = f(t-3)u_3(t)$ for $f(t) = \sin t$ is:

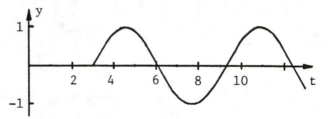

2b. In order to use Theorem 6.4 we must write f(t) in terms

    of $u_c(t)$. Since $t^2-2t + 2 = (t-1)^2 + 1$ (by completing

    the square), we can thus write $f(t) = u_1(t)g(t-1)$, where

    $g(t) = t^2+1$. Now applying Theorem 6.4 we have

$$\mathcal{L}\{f(t)\} = \mathcal{L}\{u_1(t)g(t-1)\} = e^{-s}\mathcal{L}\{g(t)\} = e^{-s}[(2/s^3) +$$

$(1/s)]$.

3b. Use partial fractions to write $F(s) = e^{-2s}[(s-1)^{-1} -$
(s+2)^{-1}]/3$. For ease in calculations let us define $G(s)$
$= (s-1)^{-1}$ and $H(s) = (s+2)^{-1}$. Then $F(s) = [e^{-2s}G(s) -$
$e^{-2s}H(s)]/3$. Using the fact that $\mathcal{L}\{e^{at}\} = (s-a)^{-1}$ and
applying Theorem 6.4, we have $F(s) = [e^{-2s}\mathcal{L}\{e^t\} - e^{-2s} \cdot$
$\mathcal{L}\{e^{-2t}\}]/3$. Thus, $F(s) = [\mathcal{L}\{u_2(t)e^{(t-2)}\} -$
$\mathcal{L}\{u_2(t)e^{-2(t-2)}\}]/3$. Using the linearity of the Laplace
transform, we have $\mathcal{L}\{f(t)\} = \mathcal{L}\{u_2(t)[e^{t-2} - e^{-2(t-2)}]/3\}$.
Hence, $f(t) = [u_2(t)(e^{t-2} - e^{-2(t-2)})]/3$. An alternate
method is to complete the square in the denominator:
$$F(s) = \frac{e^{-2s}}{(s+1/2)^2 - 9/4} .$$ This gives $f(t) =$
$(2/3)u_2(t)e^{-(t-2)/2} \sinh \frac{3}{2} (t-2)$, which is the same as
that found above.

5b. By completing the square in the denominator of F we can
write $F(s) = (2s+1)/[(2s+1)^2 + 4]$. This has the form
$G(2s+1)$ where $G(u) = u/(u^2+4)$. We must find
$\mathcal{L}^{-1}\{G(2s+1)\}$. Applying the results of Problem 4(c), we
have $\mathcal{L}^{-1}\{F(s)\} = \frac{1}{2} e^{-t/2} \cos (\frac{2t}{2})$.

5c. If the approach of Problem 5b is used we find $f(t) =$
$(1/3)e^{2t/3} \sinh(t/3)$, which is equivalent to the correct
answer using the definition of sinh t.

6d. Assuming that term-by-term integration of the infinite

series is permissible and recalling that $\mathcal{L}\{u_c(t)\} =$

e^{-cs}/s for $s > 0$, we have $\mathcal{L}\{f(t)\} = (1/s) + \sum_{k=1}^{\infty} (-1)^k$.

$\mathcal{L}\{u_k(t)\} = (1/s) + \sum_{k=1}^{\infty} (-1)^k e^{-ks}/s = [\sum_{k=0}^{\infty} (-e^{-s})^k]/s.$ We

recognize the last infinite series as the geometric

series, $\sum_{k=0}^{\infty} ar^k$, with $r = -e^{-s}$. This series converges to

$[1/(1+e^{-s})]$ if $|r| < 1$. Hence,

$\mathcal{L}\{f(t)\} = (1/s)[1/(1+e^{-s})]$, $s > 0$.

7. Using the definition of the Laplace transform we have

$F(s) = \mathcal{L}\{f(t)\} = \int_0^{\infty} e^{-st} f(t)dt.$ Since f is periodic with

period T, we have $f(t+T) = f(t)$. This suggests that we

rewrite the improper integral as $\int_0^{\infty} e^{-st} f(t)dt =$

$\sum_{n=0}^{\infty} \int_{nT}^{(n+1)T} e^{-st} f(t)dt.$ The periodicity of f also

suggests that we make the change of variable $t = r + nT$.

Hence, $F(s) = \sum_{n=0}^{\infty} \int_0^T e^{-s(r+nT)} f(r+nT)dr = \sum_{n=0}^{\infty} (e^{-sT})^n \cdot$

$\int_0^T e^{-rs} f(r)dr$, where we have used the fact that $f(r+nT)$

$= f(r+(n-1)T) = \ldots = f(r+T) = f(r)$ from the definition

that f is periodic. We recognize this last series as the

geometric series, $\sum_{n=0}^{\infty} au^n$, with $a = \int_0^T e^{-rs} f(r)dr$ and

$u = e^{-sT}$. The geometric series converges to $a/(1-u)$ for

$|u| < 1$ and consequently we obtain

$F(s) = (1 - e^{-sT})^{-1} \int_0^T e^{-rs} f(r)dr$, $s > 0$.

8b.   The function f is periodic with period 2. The result of

Problem 7 gives us $\mathcal{L}\{f(t)\} = \int_0^2 e^{-st} f(t)dt/(1-e^{-2s})$.

Calculating the integral we have $\int_0^2 e^{-st} f(t)dt =$

$\int_0^1 e^{-st} dt - \int_1^2 e^{-st}dt = (1-e^{-s})/s + (e^{-2s}-e^{-s})/s =$

$(e^{-2s}-2e^{-s}+1)/s = (1-e^{-s})^2/s$. Since the denominator of

$\mathcal{L}\{f(t)\}$, $1 - e^{-2s}$, may be written as $(1-e^{-s})(1+e^{-s})$ we

obtain the desired answer.

Section 6.3.1, Page 309

1.   f(t) can be written in the form $f(t) = 1 - u_{\pi/2}(t)$ and

thus the Laplace transform of the D.E. is $(s^2+1)Y(s) -$

$sy(0) - y'(0) = (1/s) - e^{-\pi s/2}/s$. Introducing the I.C.

and solving for Y(s), we obtain $Y(s) = (s^2+1)^{-1} +$

$[s(s^2+1)]^{-1} - e^{-\pi s/2}/s(s^2+1)$. Using partial fractions

on the second and third term we find $Y(s) = (1/s) +$

$(s^2+1)^{-1} - s/(s^2+1) - e^{-\pi s/2}/s + e^{-\pi s/2} s/(s^2+1)$. The

inverse transform of the first three terms can be

obtained directly from Table 6.1. Using Theorem 6.4 to

find the inverse transform of the last two terms we have

$\mathcal{L}^{-1}\{e^{-s/2}/s\} = u_{\pi/2}(t)g(t - \pi/2)$ where $g(t) = \mathcal{L}^{-1}\{1/s\} =$

1 and $\mathcal{L}^{-1}\{e^{-s/2} s/(s^2+1)\} = u_{\pi/2}(t)h(t - \pi/2)$ where

$h(t) = \mathcal{L}^{-1}\{s/(s^2+1)\} = \cos t$. Hence, $y = 1 + \sin t -$

$\cos t + u_{\pi/2}(t)[\cos(t - \pi/2) - 1] = 1 + \sin t - \cos t -$

$u_{\pi/2}(t)[1 - \sin t]$.

3.   According to Theorem 6.4, $\mathcal{L}\{u_{2\pi}(t)\sin(t-2\pi)\} = e^{-2\pi s}$.

$\mathcal{L}\{\sin t\} = e^{-2\pi s}/(s^2+1)$.   Transforming the D.E., we have

$(s^2+4)Y(s) - sy(0) - y'(0) = 1/(s^2+1) - e^{-2\pi s}/(s^2+1)$.

Introducing the I.C. and solving for Y(s), we obtain Y(s)

$= (1-e^{-2\pi s})/(s^2+1)(s^2+4)$.   We apply partial fractions to

write $Y(s) = [(s^2+1)^{-1} - (s^2+4)^{-1} - e^{-2\pi s}(s^2+1)^{-1} +$

$e^{-2\pi s}(s^2+4)^{-1}]/3$.   We compute the inverse transform of

the first two terms directly from Table 6.1 after noting

that $(s^2+4)^{-1} = (1/2)[2/(s^2+4)]$.   We apply Theorem 6.4 to

the last two terms to obtain the solution, y =

(1/3) $\{\sin t - (1/2)\sin 2t - u_{2\pi}(t)[\sin(t-2\pi) -$

$(1/2)\sin 2(t-2\pi)]\}$.   This may be simplified using

trigonometric identities to

$$y = [(2 \sin t - \sin 2t)(1 - u_{2\pi}(t))]/6.$$

8.   Taking the Laplace transform, applying the I.C. and using

Theorem 6.4 (or referring to Example 2) we have

$(s^2+s+5/4)Y(s) = (1-e^{-\pi s/2})/s^2$.   Thus $Y(s) = \dfrac{1-e^{-s/2}}{s^2(s^2+s+5/4)}$

$= (1-e^{-\pi s/2})\left\{\dfrac{4/5}{s^2} - \dfrac{16/25}{s} + \dfrac{(16/25)s-4/25}{(s+1/2)^2+1}\right\} = (1-e^{-\pi s/2})\cdot$

H(s), where we have used partial fractions and completed

the square in the denominator of the last term.   Since

the numerator of the last term of H can be written as

$\dfrac{16}{25}[(s+1/2) - 3/4]$, we see that $\mathcal{L}^{-1}\{H(s)\} = (4/25)(5t -4$

$+ 4e^{-t/2} \cos t - 3e^{-t/2} \sin t$), which yields the desired

solution using Theorem 6.4.

10. Note that $g(t) = \sin t - u_\pi(t) \sin t = \sin t + u_\pi(t) \cdot$

$\sin(t-\pi)$. Proceeding as in Problem 8 we find $Y(s) =$

$(1+e^{-\pi s})\dfrac{1}{(s^2+1)(s^2+s+5/4)}$. The correct partial fraction

expansion of the quotient is $\dfrac{as+b}{s^2+1} + \dfrac{cs+d}{s^2+s+5/4}$, where $a+c =$

0, $a+b+d = 0$, $(5/4)a+b+c = 0$ and $(5/4)b+d = 1$ by equating

coefficients. Solving for the constants yields the

desired solution.

14. Since f is periodic with period 2π, we can apply the

result of Problem 7 in Section 6.3 to obtain $\mathcal{L}\{f(t)\} =$

$\int_0^{2\pi} e^{-st}f(t)dt/(1 - e^{-2\pi s}) = 1/s(1 + e^{-\pi s})$. Thus, the

transformed equation is $(s^2+1)Y(s) - sy(0) - y'(0) = [s(1$

$+ e^{-\pi s})]^{-1}$. Introducing the I.C. and solving gives

$Y(s) = \dfrac{s}{s^2+1} + \dfrac{1}{(s^2+1)s(1+e^{-s})} = \dfrac{s}{s^2+1} + \dfrac{1}{(1+e^{-s})}[\dfrac{1}{s} - \dfrac{s}{s^2+1}]$

where partial fractions have been used to get the terms

in square brackets. The inverse transform of this,

however, cannot be found in Table 6.1. Recall that

Problem 6(d), Section 6.3, gave a result that had $(1 +$

$e^{-s})$ in the denominator. This indicates that we should

attempt to write the second term in the expression for

$Y(s)$ as a series. In this case we can write $[1 + e^{-\pi s}]^{-1}$

as the sum of the geometric series, $[1 + e^{-\pi s}]^{-1} =$

$\sum\limits_{n=0}^{\infty} (-e^{-\pi s})^n = 1 + \sum\limits_{n=1}^{\infty} (-e^{-\pi s})^n.$ Thus $Y(s) = s/(s^2+1) + [1+$

$\sum\limits_{n=1}^{\infty} (-e^{-\pi s})^n][(1/s) - s/(s^2+1)] = (1/s) + \sum\limits_{n=1}^{\infty} (-1)^n e^{-n\pi s}.$

$[(1/s) - s/(s^2+1)].$ Assuming that term-by-term inversion

of the infinite series is permissible, we apply Theorem

6.4 and Table 6.1 to obtain $y = 1 + \sum\limits_{n=1}^{\infty} (-1)^n u_{n\pi}(t)[1 -$

$\cos(t-n\pi)].$

Section 6.4, Page 314

1. Proceeding as in the Example, we take the Laplace

    transform of the D.E. and apply the I.C.: $(s^2 + 2s +$

    $2)Y(s) = s + 2 + e^{-\pi s}.$ Thus, $Y(s) = (s+2)/[(s+1)^2 + 1] +$

    $e^{-\pi s}/[(s+1)^2 + 1].$ We write the first term as

    $(s+1)/[(s+1)^2 + 1] + 1/[(s+1)^2 + 1].$ Applying Theorem

    6.4 and using Table 6.1, we obtain the solution, $y = e^{-t} \cdot$

    $\cos t + e^{-t} \sin t - u_\pi(t)e^{(t-\pi)} \sin t.$ Note that

    $\sin(t-\pi) = - \sin t.$

3. Taking the Laplace transform of the D.E. and applying the

    I.C., we obtain $(s^2 + 2s + 1)Y(s) = 2 + e^{-2\pi s}/s$ where

    $\mathcal{L}\{\delta(t)\} = 1.$ Solving for $Y(s)$ and using partial

    fractions on the last term, we have $Y(s) = 2/(s+1)^2 +$

    $e^{-2\pi s}[(1/s) - 1/(s+1) - 1/(s+1)^2].$ The inverse transform

    of $2/(s+1)^2$ can be found in Table 6.1 or by using the

    result of Problem 28 of Section 6.2 Hence, by Theorem

    6.4, we have $y = \mathcal{L}^{-1}\{Y(s)\} = 2te^{-t} + u_{2\pi}(t)[1 - e^{-(t-2\pi)}$

$- (t-2\pi)e^{-(t-2\pi)}]$.

5. Following the procedure of earlier problems we have Y(s)

    $= (s^2+2)/[(s^2+1)(s^2+2s+3)] + e^{-\pi s}/(s^2+2s+3)$. The partial

    fraction expansion of the first term is $(as+b)/(s^2+1)$ +

    $(cs+d)/(s^2+2s+3)$, where a + c = 0, 2a + b + d = 1, 3a +

    2b + c = 0 and 3b + d = 2.

7. Taking the Laplace transform of the D.E. yields

    $(s^2+1)Y(s) - y'(0) = \int_0^t e^{-st} \delta(t-\pi)\cos t\ dt$. Since $\delta(t-\pi)$

    = 0 for t ≠ π the integral on the right is equal to

    $\int_{-\infty}^{\infty} e^{-st} \delta(t-\pi)\cos t\ dt$ which equals $e^{-\pi s}\cos \pi$ from

    Eq.(16). Substituting for y'(0) and solving for Y(s)

    gives the desired solution.

13b. Substituting for f(t) we have

    $y = \int_0^t e^{-(t-\tau)} \delta(\tau-\pi)\sin(t-\tau)d\tau$. We know that the

    integration variable is always less than t (the upper

    limit) and thus for t < π we have τ<π and thus $\delta(\tau-\pi)$ = 0.

    Hence y = 0 for t < π. For t > π, utilize Eq.(16).

Section 6.5, Page 319

1c. Using the format of Eqs.(2) and (3) we have

    $f*(g*h) = \int_0^t f(t-\tau)(g*h)(\tau)d\tau$

    $= \int_0^t f(t-\tau)[\int_0^t g(\tau-\eta)h(\eta)d\eta]d\tau$

    $= \int_0^t [\int_\eta^t f(t-\tau)g(\tau-\eta)d\tau]h(\eta)d\eta$.

This last double integral is obtained from the previous line by interchanging the order of the η and τ integration. Making the change of variable $\omega = \tau - \eta$ on the inside integral and comparing the result with (f*g)*h yields the desired result.

4a. It is possible to determine f(t) explicitly by using integration by parts and then to find its transform F(s). However, it is much more convenient to apply Theorem 6.6. Let us define $g(t) = t^2$ and $h(t) = \cos 2t$. Then, $f(t) = \int_0^t g(t-\tau)h(\tau)d\tau$. Using Table 6.1, we have $G(s) = \mathcal{L}\{g(t)\} = 2/s^3$ and $H(s) = \mathcal{L}\{h(t)\} = s/(s^2+4)$. Hence, by Theorem 6.6, $\mathcal{L}\{f(t)\} = F(s) = G(s)H(s) = 2/s^2(s^2+4)$.

5a. As was done in Example 1 think of F(s) as the product of s^{-4} and $(s^2+1)^{-1}$ which, according to Table 6.1, are the transforms of $t^3/6$ and $\sin t$, respectively. Hence, by Theorem 6.6, the inverse transform of F(s) is $f(t) = (1/6) \int_0^t (t-\tau)^3 \sin \tau \, d\tau$.

7. We take the Laplace transform of the D.E. and apply the I.C., $(s^2 + 2s + 2)Y(s) = \alpha/(s^2 + \alpha^2)$. Solving for Y(s), we have $Y(s) = [\alpha/(s^2+\alpha^2)][(s+1)^2 + 1]^{-1}$, where the second factor has been written in a convenient way by completing the square. Thus Y(s) is seen to be the product of the transforms of $\sin \alpha t$ and $e^{-t} \sin t$ respectively. Hence, according to Theorem 6.6,

$$y = \int_0^t e^{-(t-\tau)} \sin(t-\tau)\sin\alpha\tau \, d\tau.$$

9. Proceeding as in the above problems we obtain

$$Y(s) = \frac{s}{s^2+s+5/4} + \frac{1-e^{-s}}{s(s^2+s+5/4)}$$

$$= \frac{(s+1/2)-1/2}{(s+1/2)^2+1} + \frac{1-e^{-s}}{s} \cdot \frac{1}{(s+1/2)^2+1},$$

where the first term is obtained by completing the square

in the denominator and the second term is written as the

product of two terms whose inverse transforms are known

so that Theorem 6.6 can be used. Note that

$\mathcal{L}^{-1}\{(1-e^{-s})/s\} = 1 - u_\pi(t)$. Also note that a different

form of the same solution would be obtained by writing

the second term as $(1-e^{-\pi s})(\frac{a}{s} + \frac{bs+c}{(s+1/2)^2+1})$ and solving

for a, b and c. In this case $\mathcal{L}^{-1}\{1-e^{-s}\} = \delta(t) -$

$\delta(t-\pi)$ from Section 6.4.

11. Taking the Laplace transform, using the I.C. and solving

for Y(s) we have Y(s) = (s+3)/(s+1)(s+2) +

$s/(s^2+\alpha^2)(s+1)(s+2)$. As in Problem 9 there are several

correct ways the second term can be treated in order to

use the convolution integral. In order to obtain the

desired answer, write the second term as $\frac{s}{s^2+\alpha^2}(\frac{a}{s+1} + \frac{b}{s+2})$

and solve for a and b.

14. To find $\Phi(s)$ you must recognize the integral that

appears in the equation as a convolution integral.

CHAPTER 7

Section 7.1, Page 328

1a. Let $x_4 = x_1'$, $x_5 = x_2'$ and $x_6 = x_3'$. Then the desired
system is $x_1' = x_4$, $x_2' = x_5$, $x_3' = x_6$,
$x_4' = F_1(t,x_1,x_2,x_3,x_4,x_5,x_6)/m$, $x_5' = F_2(t,x_1,\ldots,x_6)/m$
and $x_6' = F_3(t,x_1,\ldots,x_6)/m$.

2. Let $x_1 = u$ and $x_2 = u'$; then $x_1' = x_2$ is the first of the
desired pair of equations. The second equation is
obtained by substituting $u'' = x_2'$, $u' = x_2$, and $u = x_1$ in
the given D.E. The I.C. become $x_1(0) = u_0$, $x_2(0) = u_0'$.

3. If $a_{12} \neq 0$, then solve the first equation for x_2,
obtaining $x_2 = [x_1' - a_{11}x_1 - g_1(t)]/a_{12}$. Upon
substituting this expression into the second equation, we
have a second order linear O.D.E. for x_1. One I.C. is
$x_1(0) = x_1{}^0$. The second I.C. is $x_2(0) = [x_1'(0) - a_{11}x_1(0) - g_1(0)]/a_{12} = x_2{}^0$. Solving for $x_1'(0)$ gives
$x_1'(0) = a_{12}x_2{}^0 + a_{11}x_1{}^0 + g_1(0)$. These results hold
when $a_{11},\ldots,a_{22}$ are functions of t as long as the
derivatives exist and $a_{12}(t)$ and $a_{21}(t)$ are not both zero
on the interval. The initial conditions will involve
$a_{11}(0)$ and $a_{12}(0)$.

8. Let us number the nodes 1,2, and 3 clockwise beginning
with the top right node in Figure 7.4. Also let I_1, I_2,

I_3, and I_4 denote the currents through the resistor $R = 1$, the inductor $L = 1$, the capacitor $C = \frac{1}{2}$, and the resistor $R = 2$, respectively. Let V_1, V_2, V_3, and V_4 be the corresponding voltage drops. Kirchhoff's first law applied to nodes 1 and 2, respectively, gives (i) $I_1 - I_2 = 0$ and (ii) $I_2 - I_3 - I_4 = 0$. Kirchhoff's second law applied to each loop gives (iii) $V_1 + V_2 + V_3 = 0$ and (iv) $V_3 - V_4 = 0$. The current–voltage relation through each circuit element yields four more equations: (v) $V_1 = I_1$, (vi) $I_2' = V_2$, (vii) $(1/2)V_3' = I_3$ and (viii) $V_4 = 2I_4$. We thus have a system of eight equations in eight unknowns, and we wish to eliminate all of the variables except I_2 and V_3 from this system of equations. For example, we can use Eqs.(i) and (iv) to eliminate I_1 and V_4 in Eqs. (v) and (viii). Then use the new Eqs. (v) and (viii) to eliminate V_1 and I_4 in Eqs. (ii) and (iii). Finally, use the new Eqs. (ii) and (iii) in Eqs. (vi) and (vii) to obtain $I_2' = -I_2 - V_3$, $V_3' = 2I_2 - V_3$. These equations are identical (when subscripts on the remaining variables are dropped) to the equations given in the text.

Section 7.2, Page 338

1c.   Using Eq.(9) and following Example 1 we have

$$AB = \begin{pmatrix} 4 + 2 + 0 & -2 + 10 + 0 & 3 + 0 + 0 \\ 12 - 2 - 6 & -6 + 10 - 1 & 9 + 0 - 2 \\ -8 - 1 + 18 & 4 + 5 + 3 & -6 + 0 + 6 \end{pmatrix},$$

which yields the correct answer.

In problems 10 through 19 the method of row reduction illustrated in Example 2 can be used to find the inverse matrix or else to show that none exists. We start with the original matrix augmented by the identity matrix, describe a suitable sequence of elementary row operations, and show the result of applying these operations.

10. Start with the given matrix augmented by the identity matrix.

$$\left(\begin{array}{cc:cc} 1 & 4 & 1 & 0 \\ -2 & 3 & 0 & 1 \end{array} \right)$$

Add 2 times the first row to the second row.

$$\left(\begin{array}{cc:cc} 1 & 4 & 1 & 0 \\ 0 & 11 & 2 & 1 \end{array} \right)$$

Multiply the second row by (1/11)

$$\left(\begin{array}{cc:cc} 1 & 4 & 1 & 0 \\ 0 & 1 & 2/11 & 1/11 \end{array} \right)$$

Add (-4) times the second row to the first row.

$$\left(\begin{array}{cc:cc} 1 & 0 & 3/11 & -4/11 \\ 0 & 1 & 2/11 & 1/11 \end{array} \right)$$

The 2 x 2 matric appearing on the right side of this augmented matrix is the desired inverse matrix. The answer can be checked by multiplying it by the given matrix; the result should be the identity matrix.

12. The augmented matrix in this case is:

$$\left(\begin{array}{ccc:ccc} 1 & 2 & 3 & 1 & 0 & 0 \\ 2 & 4 & 5 & 0 & 1 & 0 \\ 3 & 5 & 6 & 0 & 0 & 1 \end{array} \right)$$

Add (−2) times the first row to the second row and (−3) times the first row to the third row.

$$\begin{pmatrix} 1 & 2 & 3 & \vdots & 1 & 0 & 0 \\ 0 & 0 & -1 & \vdots & -2 & 1 & 0 \\ 0 & -1 & -3 & \vdots & -3 & 0 & 1 \end{pmatrix}$$

Multiply the second and third rows by (−1) and interchange them.

$$\begin{pmatrix} 1 & 2 & 3 & \vdots & 1 & 0 & 0 \\ 0 & 1 & 3 & \vdots & 3 & 0 & -1 \\ 0 & 0 & 1 & \vdots & 2 & -1 & 0 \end{pmatrix}$$

Add (−3) times the third row to the first and second rows.

$$\begin{pmatrix} 1 & 2 & 0 & \vdots & -5 & 3 & 0 \\ 0 & 1 & 0 & \vdots & -3 & 3 & -1 \\ 0 & 0 & 1 & \vdots & 2 & -1 & 0 \end{pmatrix}$$

Add (−2) times the second row to the first row.

$$\begin{pmatrix} 1 & 0 & 0 & \vdots & 1 & -3 & 2 \\ 0 & 1 & 0 & \vdots & -3 & 3 & -1 \\ 0 & 0 & 1 & \vdots & 2 & -1 & 0 \end{pmatrix}$$

The desired answer appears on the right side of this augmented matrix.

14. Again, start with the given matrix augmented by the identity matrix.

$$\begin{pmatrix} 1 & 2 & 1 & \vdots & 1 & 0 & 0 \\ -2 & 1 & 8 & \vdots & 0 & 1 & 0 \\ 1 & -2 & -7 & \vdots & 0 & 0 & 1 \end{pmatrix}$$

Add (2) times the first row to the second row and add (−1) times the first row to the third row.

$$\begin{pmatrix} 1 & 2 & 1 & \vdots & 1 & 0 & 0 \\ 0 & 5 & 10 & \vdots & 2 & 1 & 0 \\ 0 & -4 & -8 & \vdots & -1 & 0 & 1 \end{pmatrix}$$

Multiply the second row by (1/5).
$$\begin{pmatrix} 1 & 2 & 1 & . & 1 & 0 & 0 \\ & & & . & & & \\ 0 & 1 & 2 & . & 2/5 & 1/5 & 0 \\ & & & . & & & \\ 0 & -4 & -8 & . & -1 & 0 & 1 \end{pmatrix}$$

Add (-2) times the second row to the first row and add (4) times the second row to the third row.
$$\begin{pmatrix} 1 & 0 & -3 & . & 1/5 & -2/5 & 0 \\ & & & . & & & \\ 0 & 1 & 2 & . & 2/5 & 1/5 & 0 \\ & & & . & & & \\ 0 & 0 & 0 & . & 3/5 & 4/5 & 1 \end{pmatrix}$$

Since the element in the third row and third column is zero, no further reduction can be performed. Since it cannot be reduced to the identity matrix, the given matrix is singular.

23. $\underset{\sim}{x}' = \begin{pmatrix} 4 \\ 2 \end{pmatrix} 2e^{2t} = \begin{pmatrix} 8 \\ 4 \end{pmatrix} e^{2t}$;

$$\begin{pmatrix} 3 & -2 \\ 2 & -2 \end{pmatrix} \underset{\sim}{x} = \begin{pmatrix} 3 & -2 \\ 2 & -2 \end{pmatrix} \begin{pmatrix} 4 \\ 2 \end{pmatrix} e^{2t} = \begin{pmatrix} 12-4 \\ 8-4 \end{pmatrix} e^{2t} = \begin{pmatrix} 8 \\ 4 \end{pmatrix} e^{2t}$$

Section 7.3, Page 350

1. Form the augmented matrix, as in Example 1, and use row reduction.
$$\begin{pmatrix} 1 & 0 & -1 & . & 0 \\ & & & . & \\ 3 & 1 & 1 & . & 1 \\ & & & . & \\ -1 & 1 & 2 & . & 2 \end{pmatrix}$$

Add (-3) times the first row to the second and add the first row to the third
$$\begin{pmatrix} 1 & 0 & -1 & . & 0 \\ & & & . & \\ 0 & 1 & 4 & . & 1 \\ & & & . & \\ 0 & 1 & 1 & . & 2 \end{pmatrix}$$

Add (-1) times the second row to the third.
$$\begin{pmatrix} 1 & 0 & -1 & . & 0 \\ & & & . & \\ 0 & 1 & 4 & . & 1 \\ & & & . & \\ 0 & 0 & -3 & . & 1 \end{pmatrix}$$

The third row is equivalent to $-3x_3 = 1$ or $x_3 = -1/3$.

Likewise the second row is equivalent to $x_2 + 4x_3 = 1$, so

$x_2 = 7/3$. Finally, from the first row, $x_1 - x_3 = 0$, so

$x_1 = -1/3$. The answer $x_1 = -1/3$, $x_2 = 7/3$ and $x_3 = $

$-1/3$, of course, could be checked by substituting into

the original equations.

2. Form the augmented matrix and
    use row reduction to obtain

$$\left(\begin{array}{ccc:c} 1 & 2 & -1 & 1 \\ 0 & -3 & 3 & -1 \\ 0 & 0 & 0 & 1 \end{array} \right).$$

The last row corresponds to the equation $0x_1 + 0x_2 + 0x_3$

$= 1$, and there is no choice of x_1, x_2, and x_3 that

satisfies this equation. Hence the given system of

equations has no solution.

3. Form the augmented matrix and
    use row reduction.

$$\left(\begin{array}{ccc:c} 1 & 2 & -1 & 2 \\ 2 & 1 & 1 & 1 \\ 1 & -1 & 2 & -1 \end{array} \right)$$

Add (-2) times the first row to
the second and add (-1) times the
first row to the third.

$$\left(\begin{array}{ccc:c} 1 & 2 & -1 & 2 \\ 0 & -3 & 3 & -3 \\ 0 & -3 & 3 & -3 \end{array} \right)$$

Add (-1) times the second row to
the third row and then multiply
the second row by $(-1/3)$.

$$\left(\begin{array}{ccc:c} 1 & 2 & -1 & 2 \\ 0 & 1 & -1 & 1 \\ 0 & 0 & 0 & 0 \end{array} \right)$$

Since the last row has only zero entries, it may be

dropped. The second row corresponds to the equation x_2 –

$x_3 = 1$. We can assign an arbitrary value to either x_2 or

x_3 and use this equation to solve for the other. For

example, let $x_3 = c$, where c is arbitrary. Then $x_2 = 1 +$

c. The first row corresponds to the equation $x_1 + 2x_2$ –

$x_3 = 2$, so $x_1 = 2 - 2x_2 + x_3 = 2 - 2(1+c) + c = -c$.

6a. To determine whether the given set of vectors is linearly

independent we must solve the system $c_1 x^{(1)} + c_2 x^{(2)} +$

$c_3 x^{(3)} = 0$ for c_1, c_2, and c_3.

Form the augmented matrix and
use row reduction.

$$\begin{pmatrix} 1 & 0 & 1 & . & 0 \\ 1 & 1 & 0 & . & 0 \\ 0 & 1 & 1 & . & 0 \end{pmatrix}$$

Add (–1) times the first row
to the second.

$$\begin{pmatrix} 1 & 0 & 1 & . & 0 \\ 0 & 1 & -1 & . & 0 \\ 0 & 1 & 1 & . & 0 \end{pmatrix}$$

Add (–1) times the second row
to the third.

$$\begin{pmatrix} 1 & 0 & 1 & . & 0 \\ 0 & 1 & -1 & . & 0 \\ 0 & 0 & 2 & . & 0 \end{pmatrix}$$

From the third row we have $c_3 = 0$. Then from the second

row, $c_2 - c_3 = 0$, so $c_2 = 0$. Finally from the first row

$c_1 + c_3 = 0$, so $c_1 = 0$. Since $c_1 = c_2 = c_3 = 0$, we

conclude that the given vectors are linearly independent.

6c. As in Problem 6a we wish to solve the system $c_1 x^{(1)} +$

$$c_2 \underset{\sim}{x}^{(2)} + c_3 \underset{\sim}{x}^{(3)} + c_4 \underset{\sim}{x}^{(4)} = \underset{\sim}{0} \text{ for } c_1, c_2, c_3, \text{ and } c_4.$$

Form the augmented matrix and use row reduction.

$$\begin{pmatrix} 1 & -1 & -2 & -3 & . & 0 \\ 2 & 0 & -1 & 0 & . & 0 \\ 2 & 3 & 1 & -1 & . & 0 \\ 3 & 1 & 0 & 3 & . & 0 \end{pmatrix}$$

Add (-2) times the first row to the second, add (-2) times the first row to the third, and add (-3) times the first row to the fourth.

$$\begin{pmatrix} 1 & -1 & -2 & -3 & . & 0 \\ 0 & 2 & 3 & 6 & . & 0 \\ 0 & 5 & 5 & 5 & . & 0 \\ 0 & 4 & 6 & 12 & . & 0 \end{pmatrix}$$

Multiply the second row by $(1/2)$ and then add (-5) times the second row to the third and add (-4) times the second row to the fourth.

$$\begin{pmatrix} 1 & -1 & -2 & -3 & . & 0 \\ 0 & 1 & 3/2 & 3 & . & 0 \\ 0 & 0 & -5/2 & -10 & . & 0 \\ 0 & 0 & 0 & 0 & . & 0 \end{pmatrix}$$

The third row is equivalent to the equation $c_3 + 4c_4 = 0$.

One way to satisfy this equation is by choosing $c_4 = -1$; then $c_3 = 4$. From the second row we have $c_2 = -(3/2)c_3$ $- 3c_4 = -6 + 3 = -3$. Then, from the first row, $c_1 = c_2$ $+ 2c_3 + 3c_4 = -3 + 8 - 3 = 2$. Hence the given vectors are linearly dependent, and satisfy $2\underset{\sim}{x}^{(1)} - 3\underset{\sim}{x}^{(2)} + 4\underset{\sim}{x}^{(3)}$ $- \underset{\sim}{x}^{(4)} = \underset{\sim}{0}$.

9. Let $t = t_0$ be a fixed value of t in the interval $0 \le t \le 1$. To determine whether $\underset{\sim}{x}^{(1)}(t_0)$ and $\underset{\sim}{x}^{(2)}(t_0)$ are linearly dependent we must solve $c_1\underset{\sim}{x}^{(1)}(t_0) + c_2\underset{\sim}{x}^{(2)}(t_0) = \underset{\sim}{0}$. We

have the augmented matrix

$$
\begin{pmatrix}
e^{t_0} & 1 & \cdot & 0 \\
 & & \cdot & \\
 & & \cdot & \\
t_0 e^{t_0} & t_0 & \cdot & 0
\end{pmatrix}
$$

and by row reduction we obtain

$$
\begin{pmatrix}
e^{t_0} & 1 & \cdot & 0 \\
 & & \cdot & \\
 & & \cdot & \\
0 & 0 & \cdot & 0
\end{pmatrix}.
$$

Thus, for example, we can choose $c_1 = 1$ and $c_2 = -e^{t_0}$, and

hence the given vectors are linearly dependent at t_0.

Since t_0 is arbitrary the vectors are linearly dependent

at each point in the interval. However, there is no

linear relation between $\underset{\sim}{x}^{(1)}$ and $\underset{\sim}{x}^{(2)}$ that is valid

throughout the interval $0 \leq t \leq 1$. For example, if $t_1 \neq$

t_0, and if c_1 and c_2 are chosen as above, then

$c_1 \underset{\sim}{x}^{(1)}(t_1) + c_2 \underset{\sim}{x}^{(2)}(t_1)$

$$
= \begin{pmatrix} e^{t_1} \\ t_1 e^{t_1} \end{pmatrix} + -e^{t_0}\begin{pmatrix} 1 \\ t_1 \end{pmatrix} = \begin{pmatrix} e^{t_1} - e^{t_0} \\ t_1 e^{t_1} - t_1 e^{t_0} \end{pmatrix} \neq \begin{pmatrix} 0 \\ 0 \end{pmatrix}.
$$

Hence the given vectors must be linearly independent on

$0 \leq t \leq 1$. In fact, the same argument applies to any

interval.

10. To find the eigenvalues and eigenvectors of the given

matrix we must solve $\begin{pmatrix} 5-\lambda & -1 \\ 3 & 1-\lambda \end{pmatrix}\begin{pmatrix} x_1 \\ x_2 \end{pmatrix} = \begin{pmatrix} 0 \\ 0 \end{pmatrix}$. The determinant of coefficients is $(5-\lambda)(1-\lambda) - (-1)(3) = 0$,

or $\lambda^2 - 6\lambda + 8 = 0$. Hence $\lambda_1 = 2$ and $\lambda_2 = 4$ are the

eigenvalues. The eigenvector corresponding to λ_1, must

satisfy $\begin{pmatrix} 3 & -1 \\ 3 & -1 \end{pmatrix} \begin{pmatrix} x_1 \\ x_2 \end{pmatrix} = \begin{pmatrix} 0 \\ 0 \end{pmatrix}$, or $3x_1 - x_2 = 0$. If

we let $x_1 = 1$, then $x_2 = 3$ and the eigenvector is

$\underset{\sim}{x}^{(1)} = \begin{pmatrix} 1 \\ 3 \end{pmatrix}$, or any constant multiple of this vector.

Similarly, the eigenvector corresponding to λ_2 must

satisfy $\begin{pmatrix} 1 & -1 \\ 3 & -3 \end{pmatrix} \begin{pmatrix} x_1 \\ x_2 \end{pmatrix} = \begin{pmatrix} 0 \\ 0 \end{pmatrix}$, or $x_1 - x_2 = 0$. Hence

$\underset{\sim}{x}^{(2)} = \begin{pmatrix} 1 \\ 1 \end{pmatrix}$, or a multiple thereof.

13. The given matrix is Hermitian so we know in advance that

its eigenvalues are real. To find the eigenvalues and

eigenvectors we must solve $\begin{pmatrix} 1-\lambda & i \\ -i & 1-\lambda \end{pmatrix} \begin{pmatrix} x_1 \\ x_2 \end{pmatrix} = \begin{pmatrix} 0 \\ 0 \end{pmatrix}$. The

determinant of coefficients is $(1-\lambda)^2 - i(-i) = \lambda^2 - 2\lambda$,

so the eigenvalues are $\lambda_1 = 0$ and $\lambda_2 = 2$; observe that

they are indeed real even though the given matrix has

imaginary entries. The eigenvector corresponding to λ_1

must satisfy $\begin{pmatrix} 1 & i \\ -i & 1 \end{pmatrix} \begin{pmatrix} x_1 \\ x_2 \end{pmatrix} = \begin{pmatrix} 0 \\ 0 \end{pmatrix}$, or $x_1 + ix_2 = 0$.

Note that the second equation $-ix_1 + x_2 = 0$ is a multiple of the first. If $x_1 = 1$, then $x_2 = i$, and the

eigenvector is $\underset{\sim}{x}^{(1)} = \begin{pmatrix} 1 \\ i \end{pmatrix}$. In a similar way we find

that the eigenvector associated with λ_2 is

$\underset{\sim}{x}^{(2)} = \begin{pmatrix} 1 \\ -1 \end{pmatrix}$.

15. The eigenvalues and eigenvectors satisfy

$\begin{pmatrix} 1-\lambda & -4 \\ 4 & -7-\lambda \end{pmatrix} \begin{pmatrix} x_1 \\ x_2 \end{pmatrix} = \begin{pmatrix} 0 \\ 0 \end{pmatrix}$. The determinant of

coefficients is $(1-\lambda)(-7-\lambda) - (-4)4 = \lambda^2 + 6\lambda + 9$, so

$\lambda_1 = \lambda_2 = -3$. Thus -3 is a double eigenvalue. The

corresponding eigenvectors satisfy

$\begin{pmatrix} 4 & -4 \\ 4 & -4 \end{pmatrix} \begin{pmatrix} x_1 \\ x_2 \end{pmatrix} = \begin{pmatrix} 0 \\ 0 \end{pmatrix}$. Hence $x_1 - x_2 = 0$, so $x_1 = x_2$

and $\underset{\sim}{x}^{(1)} = \begin{pmatrix} 1 \\ 1 \end{pmatrix}$, or any multiple thereof. There are no

other linearly independent eigenvectors in this problem

since both equations in this last set are identical.

19. Since the given matrix is real and symmetric, we know

that the eigenvalues are real. Further, even if there

are repeated eigenvalues, there will be a full set of

three linearly independent eigenvectors. To find the

eigenvalues and eigenvectors we must solve

$$\begin{pmatrix} 3-\lambda & 2 & 4 \\ 2 & -\lambda & 2 \\ 4 & 2 & 3-\lambda \end{pmatrix} \begin{pmatrix} x_1 \\ x_2 \\ x_3 \end{pmatrix} = \begin{pmatrix} 0 \\ 0 \\ 0 \end{pmatrix}.$$ The determinant of

coefficients is $(3-\lambda)[-\lambda(3-\lambda)-4] - 2[2(3-\lambda) -8] + 4[4+4\lambda]$

$= -\lambda^3 + 6\lambda^2 + 15\lambda + 8$. Setting this equal to zero and

solving we find $\lambda_1 = \lambda_2 = -1$, $\lambda_3 = 8$. The eigenvectors

corresponding to λ_1 and λ_2 must satisfy

$$\begin{pmatrix} 4 & 2 & 4 \\ 2 & 1 & 2 \\ 4 & 2 & 4 \end{pmatrix} \begin{pmatrix} x_1 \\ x_2 \\ x_3 \end{pmatrix} = \begin{pmatrix} 0 \\ 0 \\ 0 \end{pmatrix};$$ hence there is only the

single relation $2x_1 + x_2 + 2x_3 = 0$ to be satisfied.

Consequently, two of the variables can be selected

arbitrarily and the third is then determined by this

equation. For example, if $x_1 = 1$ and $x_3 = 1$, then $x_2 =$

-4, and we obtain the eigenvector $\underset{\sim}{x}^{(1)} = \begin{pmatrix} 1 \\ -4 \\ 1 \end{pmatrix}$.

Similarly, if $x_1 = 1$ and $x_2 = 0$, then $x_3 = -1$, and

we have the eigenvector $\underset{\sim}{x}^{(2)} = \begin{pmatrix} 1 \\ 0 \\ -1 \end{pmatrix}$, which is linearly

independent of $\underset{\sim}{x}^{(1)}$. There are many other choices that

could have been made; however, by Eq.(28) there can be no

more than two linearly independent eigenvectors

corresponding to the eigenvalue -1. To find the

eigenvector corresponding to λ_3 we must solve

$$\begin{pmatrix} -5 & 2 & 4 \\ 2 & -8 & 2 \\ 4 & 2 & -5 \end{pmatrix} \begin{pmatrix} x_1 \\ x_2 \\ x_3 \end{pmatrix} = \begin{pmatrix} 0 \\ 0 \\ 0 \end{pmatrix}.$$ By row reduction we can

obtain the equivalent system $x_1 - 4x_2 + x_3 = 0$, $2x_2 - x_3$
$= 0$. Since there are two equations to satisfy only one
variable can be assigned an arbitrary value. If we let
$x_2 = 1$, then $x_3 = 2$ and $x_1 = 2$, so we find that

$$\underset{\sim}{x}^{(3)} = \begin{pmatrix} 2 \\ 1 \\ 2 \end{pmatrix}.$$

20a. From the discussion leading to Eq.(44) we find $\underset{\sim}{T}$ is

composed of the eigenvectors found in Problem 10 and thus

$\underset{\sim}{T} = \begin{pmatrix} 1 & 1 \\ 3 & 1 \end{pmatrix}$. To verify Eq.(46) we find $\underset{\sim}{T}^{-1} = \begin{pmatrix} -1/2 & 1/2 \\ 3/2 & -1/2 \end{pmatrix}$

and calculate $\underset{\sim}{T}^{-1}\underset{\sim\sim}{AT}$ to find that $\underset{\sim}{T}^{-1}\underset{\sim\sim}{AT} = \begin{pmatrix} 2 & 0 \\ 0 & 4 \end{pmatrix}$, as

indicated in Eq.(45).

Section 7.4, Page 356

2a. From Eq.(10) we have

$$W = \begin{vmatrix} x_1^{(1)} & x_1^{(2)} \\ x_2^{(1)} & x_2^{(2)} \end{vmatrix} = x_1^{(1)}x_2^{(2)} -$$

$x_2^{(1)}x_1^{(2)}$. Taking the derivative of these two products

yields four terms which may be written as $\dfrac{dw}{dt} =$

$$\left[\frac{dx_1^{(1)}}{dt} x_2^{(2)} - x_2^{(1)} \frac{dx_1^{(2)}}{dt}\right] + \left[x_1^{(1)} \frac{dx_2^{(2)}}{dt} - \frac{dx_2^{(1)}}{dt} x_1^{(2)}\right].$$

The terms in the square brackets can now be recognized as the respective determinants appearing in the desired solution. A similar result was mentioned in Problem 2 of Section 5.2.

2b. If $x^{(1)}$ is substituted into Eq.(3) we have

$$\frac{dx_1^{(1)}}{dt} = p_{11} x_1^{(1)} + p_{12} x_2^{(1)}$$

$$\frac{dx_2^{(1)}}{dt} = p_{21} x_1^{(1)} + p_{22} x_2^{(1)}.$$

Substituting the first equation above and its counterpart for $x^{(2)}$ into the first determinant appearing in dW/dt and evaluating the result yields

$$p_{11} \begin{vmatrix} x_1^{(1)} & x_1^{(2)} \\ x_2^{(1)} & x_2^{(2)} \end{vmatrix} =$$

$p_{11}W$. Similarly, the second determinant in dW/dt is evaluated as $p_{22}W$, yielding the desired result.

6a. From Eq.(10) $W = \begin{vmatrix} t & t^2 \\ 1 & 2t \end{vmatrix}$.

6d. To obtain the system satisfied by $x^{(1)}$ and $x^{(2)}$ we proceed in a fashion similar to Problem 6 of Section 3.1. The general solution of the system can be written as $x = c_1 x^{(1)} + c_2 x^{(2)}$, or

$$\begin{pmatrix} x_1 \\ x_2 \end{pmatrix} = c_1 \begin{pmatrix} t \\ 1 \end{pmatrix} + c_2 \begin{pmatrix} t^2 \\ 2t \end{pmatrix}.$$

Taking the derivative we obtain

$$\begin{pmatrix} x_1' \\ x_2' \end{pmatrix} = c_1 \begin{pmatrix} 1 \\ 0 \end{pmatrix} + c_2 \begin{pmatrix} 2t \\ 2 \end{pmatrix} .$$

Solving this last system for c_1 and c_2 we find $c_1 = x_1'$

$- t x_2'$ and $c_2 = x_2'/2$. Thus

$$\begin{pmatrix} x_1 \\ x_2 \end{pmatrix} = (x_1' - tx_2') \begin{pmatrix} t \\ 1 \end{pmatrix} + \frac{x_2'}{2} \begin{pmatrix} t^2 \\ 2t \end{pmatrix} , \text{ which yields}$$

$$x_1 = tx_1' - \frac{t^2}{2} x_2'$$

$$x_2 = x_1' .$$

Writing this system in matrix form we have

$$\underset{\sim}{x} = \begin{pmatrix} t - t^2/2 \\ 1 \quad 0 \end{pmatrix} \underset{\sim}{x'} .$$

Finding the inverse of the matrix multiplying $\underset{\sim}{x'}$ yields

the desired solution.

Section 7.5, Page 365

1. Assuming that there are solutions of the form $\underset{\sim}{x} = \underset{\sim}{\xi} e^{rt}$,

    we substitute into the D.E. to find

    $$r \underset{\sim}{\xi} e^{rt} = \begin{pmatrix} 3 & -2 \\ 2 & -2 \end{pmatrix} \underset{\sim}{\xi} e^{rt}. \text{ Since } \underset{\sim}{\xi} = I\underset{\sim}{\xi} = \begin{pmatrix} 1 & 0 \\ 0 & 1 \end{pmatrix} \underset{\sim}{\xi},$$

    we can write this equation as $\begin{pmatrix} 3 & -2 \\ 2 & -2 \end{pmatrix} \underset{\sim}{\xi} - r \begin{pmatrix} 1 & 0 \\ 0 & 1 \end{pmatrix} \underset{\sim}{\xi} = \underset{\sim}{0}$

    and thus we must solve $\begin{pmatrix} 3-r & -2 \\ 2 & -2-r \end{pmatrix} \begin{pmatrix} \xi_1 \\ \xi_2 \end{pmatrix} = \begin{pmatrix} 0 \\ 0 \end{pmatrix}$ for r, ξ_1,

    ξ_2. The determinant of the coefficients is $(3-r)(-2-r)$

$+ 4 = r^2 - r - 2$, so the eigenvalues are $r = -1, 2$. The

eigenvector corresponding to $r = -1$ satisfies

$$\begin{pmatrix} 4 & -2 \\ 2 & -1 \end{pmatrix} \begin{pmatrix} \xi_1 \\ \xi_2 \end{pmatrix} = \begin{pmatrix} 0 \\ 0 \end{pmatrix},$$ which yields $2\xi_1 - \xi_2 = 0$. Thus

$$\underset{\sim}{x}^{(1)}(t) = \underset{\sim}{\xi}^{(1)} e^{-t} = \begin{pmatrix} 1 \\ 2 \end{pmatrix} e^{-t},$$ where we have set $\xi_1 = 1$.

(Any other non zero choice would also work). In a

similar fashion, for $r = 2$, we have $\begin{pmatrix} 1 & -2 \\ 2 & -4 \end{pmatrix} \begin{pmatrix} \xi_1 \\ \xi_2 \end{pmatrix} = \begin{pmatrix} 0 \\ 0 \end{pmatrix}$,

or $\xi_1 - 2\xi_2 = 0$. Hence $\underset{\sim}{x}^{(2)}(t) = \underset{\sim}{\xi}^{(2)} e^{2t} = \begin{pmatrix} 2 \\ 1 \end{pmatrix} e^{2t}$ by

setting $\xi_2 = 1$. The general solution is then $\underset{\sim}{x} =$

$c_1 \underset{\sim}{x}^{(1)}(t) + c_2 \underset{\sim}{x}^{(2)}(t)$. To sketch the trajectories we

follow the steps illustrated in Examples 1 and 2.

Setting $c_2 = 0$ we have $\underset{\sim}{x} = \begin{pmatrix} x_1 \\ x_2 \end{pmatrix} = c_1 \begin{pmatrix} 1 \\ 2 \end{pmatrix} e^{-t}$ or $x_1 = c_1 e^{-t}$

and $x_2 = 2c_1 e^{-t}$ and thus one asymptote is given by $x_2 = 2x_1$. In a similar fashion

$c_1 = 0$ gives $x_2 = (1/2)x_1$ as

a second asymptote. Since the

roots differ in sign, the

trajectories for this problem

are similar in nature to

those of Example 1.

5.   Proceeding as in Problem 1 we assume a solution of the

form $\underset{\sim}{x} = \underset{\sim}{\xi}e^{rt}$, where r, ξ_1, ξ_2 must now satisfy

$\begin{pmatrix} -2-r & 1 \\ 1 & -2-r \end{pmatrix} \begin{pmatrix} \xi_1 \\ \xi_2 \end{pmatrix} = \begin{pmatrix} 0 \\ 0 \end{pmatrix}$. Evaluating the determinant of

the coefficients set equal to zero yields $r = -1, -3$ as

the eigenvalues. For $r = -1$ we find $\xi_1 = \xi_2$ and thus

$\underset{\sim}{\xi}^{(1)} = \begin{pmatrix} 1 \\ 1 \end{pmatrix}$ and for $r = -3$ we find $\xi_2 = -\xi_1$ and hence

$\underset{\sim}{\xi}^{(2)} = \begin{pmatrix} 1 \\ -1 \end{pmatrix}$. The general solution is then $\underset{\sim}{x} = c_1 \begin{pmatrix} 1 \\ 1 \end{pmatrix} e^{-t}$

$+ c_2 \begin{pmatrix} 1 \\ -1 \end{pmatrix} e^{-3t}$. Since there are two negative eigenvalues,

we would expect the trajectories to be similar to those

of Example 2. Setting

$c_2 = 0$ and eliminating t

(as in Problem 1) we find

that $\begin{pmatrix} 1 \\ 1 \end{pmatrix} e^{-t}$ approaches the

origin along the line $x_2 = x_1$.

Similarly $\begin{pmatrix} 1 \\ -1 \end{pmatrix} e^{-3t}$ approaches

the origin along the line $x_2 = -x_1$.

7.   Again assuming $\underset{\sim}{x} = \underset{\sim}{\xi}e^{rt}$ we find that r, ξ_1, ξ_2 must

satisfy $\begin{pmatrix} 4-r & -3 \\ 8 & -6-r \end{pmatrix} \begin{pmatrix} \xi_1 \\ \xi_2 \end{pmatrix} = \begin{pmatrix} 0 \\ 0 \end{pmatrix}$. The determinant of the

coefficients set equal to zero yields $r = 0, -2$. For $r =$

0 we find $4\xi_1 = 3\xi_2$. Choosing $\xi_2 = 4$ we find $\xi_1 = 3$ and

thus $\xi^{(1)} = \begin{pmatrix} 3 \\ 4 \end{pmatrix}$. Similarly for r = -2 we have $\xi^{(2)} = \begin{pmatrix} 1 \\ 2 \end{pmatrix}$

and thus $x = c_1 \begin{pmatrix} 3 \\ 4 \end{pmatrix} + c_2 \begin{pmatrix} 1 \\ 2 \end{pmatrix} e^{-2t}$. To sketch the

trajectories, note that the general solution is

equivalent to the simultaneous equations $x_1 = 3c_1 +$

$c_2 e^{-2t}$ and $x_2 = 4c_1 + 2c_2 e^{-2t}$. Solving the first equation

for $c_2 e^{-2t}$ and substituting into the second yields $x_2 =$

$2x_1 - 2c_1$ and thus the trajectories are parallel straight

lines.

15. The eigenvalues and eigenvectors of the coefficient

matrix satisfy $\begin{pmatrix} 1-r & -1 & 4 \\ 3 & 2-r & -1 \\ 2 & 1 & -1-r \end{pmatrix} \begin{pmatrix} \xi_1 \\ \xi_2 \\ \xi_3 \end{pmatrix} = \begin{pmatrix} 0 \\ 0 \\ 0 \end{pmatrix}$.   The

determinant of coefficients set equal to zero reduces to

$r^3 - 2r^2 - 5r + 6 = 0$, so the eigenvalues are $r_1 = 1$, r_2

$= -2$, and $r_3 = 3$. The eigenvector corresponding to r_1

must satisfy $\begin{pmatrix} 0 & -1 & 4 \\ 3 & 1 & -1 \\ 2 & 1 & -2 \end{pmatrix} \begin{pmatrix} \xi_1 \\ \xi_2 \\ \xi_3 \end{pmatrix} = \begin{pmatrix} 0 \\ 0 \\ 0 \end{pmatrix}$. Using row

reduction we obtain the equivalent system $\xi_1 + \xi_3 = 0$,

$\xi_2 - 4\xi_3 = 0$. Letting $\xi_1 = 1$, it follows that $\xi_3 = -1$

and $\xi_2 = -4$, so $\xi^{(1)} = \begin{pmatrix} 1 \\ -4 \\ -1 \end{pmatrix}$. In a similar way the eigenvectors corresponding to r_2 and r_3 are found to be

$$\xi^{(2)} = \begin{pmatrix} 1 \\ -1 \\ -1 \end{pmatrix} \text{ and } \xi^{(3)} = \begin{pmatrix} 1 \\ 2 \\ 1 \end{pmatrix} , \text{ respectively. Thus the}$$

general solution of the given D.E. is

$$x = c_1 \begin{pmatrix} 1 \\ -4 \\ -1 \end{pmatrix} e^t + c_2 \begin{pmatrix} 1 \\ -1 \\ -1 \end{pmatrix} e^{-2t} + c_3 \begin{pmatrix} 1 \\ 2 \\ 1 \end{pmatrix} e^{3t} . \quad \text{Notice}$$

that the "trajectories" of this solution would lie in the

$x_1 \ x_2 \ x_3$ three dimensional space.

17. The eigenvalues and eigenvectors of the coefficient

matrix are found to be $r_1 = -1$, $\xi^{(1)} = \begin{pmatrix} 1 \\ 1 \end{pmatrix}$ and $r_2 = 3$,

$\xi^{(2)} = \begin{pmatrix} 1 \\ 5 \end{pmatrix}$. Thus the general solution of the given D.E.

is $x = c_1 \begin{pmatrix} 1 \\ 1 \end{pmatrix} e^{-t} + c_2 \begin{pmatrix} 1 \\ 5 \end{pmatrix} e^{3t}$. The I.C. yields the

system of equations $c_1 \begin{pmatrix} 1 \\ 1 \end{pmatrix} + c_2 \begin{pmatrix} 1 \\ 5 \end{pmatrix} = \begin{pmatrix} 1 \\ 3 \end{pmatrix}$. The

augmented matrix of this system is $\begin{pmatrix} 1 & 1 & . & 1 \\ & & . & \\ 1 & 5 & . & 3 \end{pmatrix}$ and by

row reduction we obtain $\begin{pmatrix} 1 & 1 & . & 1 \\ & & . & \\ 0 & 1 & . & 1/2 \end{pmatrix}$. Thus $c_2 = 1/2$ and

$c_1 = 1/2$. Substituting these values in the general

solution gives the solution of the I.V.P.

21. Substituting $x = \xi t^r$ into the D.E. we obtain $r\xi \, t^r =$

$\begin{pmatrix} 2 & -1 \\ 3 & -2 \end{pmatrix} \xi t^r$. For $t \neq 0$ this equation can be written as

$\begin{pmatrix} 2-r & -1 \\ 3 & -2-r \end{pmatrix} \begin{pmatrix} \xi_1 \\ \xi_2 \end{pmatrix} = \begin{pmatrix} 0 \\ 0 \end{pmatrix}$. The eigenvalues and

eigenvectors are $r_1 = 1$, $\xi^{(1)} = \begin{pmatrix} 1 \\ 1 \end{pmatrix}$ and $r_2 = -1$, $\xi^{(2)} =$

$\begin{pmatrix} 1 \\ 3 \end{pmatrix}$. Substituting these in the assumed form we obtain

the general solution $x = c_1 \begin{pmatrix} 1 \\ 1 \end{pmatrix} t + c_2 \begin{pmatrix} 1 \\ 3 \end{pmatrix} t^{-1}$.

28. The eigenvalues and eigenvectors of the coefficient

matrix were found in Problem 15 of Section 7.3, namely,

$r_1 = r_2 = -3$, $\xi^{(1)} = \begin{pmatrix} 1 \\ 1 \end{pmatrix}$. Thus one solution of the

given D.E. is $x^{(1)}(t) = \begin{pmatrix} 1 \\ 1 \end{pmatrix} e^{-3t}$ but there is no second

solution of the form $x = \xi e^{rt}$. To use the method of

reduction of order, as shown in Problem 27, let $x =$

$\begin{pmatrix} 1 & e^{-3t} \\ 0 & e^{-3t} \end{pmatrix} y$. Then it follows that $x' = \begin{pmatrix} 1 & e^{-3t} \\ 0 & e^{-3t} \end{pmatrix} y'$

$+ \begin{pmatrix} 0 & -3e^{-3t} \\ 0 & -3e^{-3t} \end{pmatrix} y$ and that $Ax = \begin{pmatrix} 1 & -3e^{-3t} \\ 4 & -3e^{-3t} \end{pmatrix} y$.

Substituting into the D.E., we obtain $\begin{pmatrix} 1 & e^{-3t} \\ 0 & e^{-3t} \end{pmatrix} y' =$

$\begin{pmatrix} 1 & 0 \\ 4 & 0 \end{pmatrix} y$. Multiplying both sides of this last equation

by $\begin{pmatrix} 1 & -1 \\ 0 & e^{3t} \end{pmatrix}$, the inverse of the matrix on the left, we obtain $\underset{\sim}{y}' = \begin{pmatrix} -3 & 0 \\ 4e^{3t} & 0 \end{pmatrix} \underset{\sim}{y}$. This equation has a variable

matrix, so it can't be solved by the methods of this

section. However, we can write the equation in scalar

form as $y_1' = -3y_1$ and $y_2' = 4e^{3t}y_1$. Solving the first

equation we get $y_1 = k_1 e^{-3t}$ and thus $y_2' = 4k_1$, which

yields $y_2 = 4k_1 t + k_2$. Consequently $\underset{\sim}{y} = k_1 \begin{pmatrix} e^{-3t} \\ 4t \end{pmatrix} +$

$k_2 \begin{pmatrix} 0 \\ 1 \end{pmatrix}$ and hence $\underset{\sim}{x} = \begin{pmatrix} 1 & e^{-3t} \\ 0 & e^{-3t} \end{pmatrix} \underset{\sim}{y} = \begin{pmatrix} k_1 + k_2 \\ k_2 \end{pmatrix} e^{-3t} +$

$4k_1 \begin{pmatrix} 1 \\ 1 \end{pmatrix} te^{-3t}$. By setting $k_1 = 1/4$ and $k_2 = -1/4$ we

obtain the form of the second solution given in the text.

See Problem 7 of Section 7.7 for another solution of this

problem.

Section 7.6, Page 374

1. We assume a solution of the form $\underset{\sim}{x} = \underset{\sim}{\xi} e^{rt}$ thus r and ξ

    are solutions of $\begin{pmatrix} 3-r & -2 \\ 4 & -1-r \end{pmatrix} \begin{pmatrix} \xi_1 \\ \xi_2 \end{pmatrix} = \begin{pmatrix} 0 \\ 0 \end{pmatrix}$. The determinant

    of coefficients is $(r^2-2r-3) + 8 = r^2 - 2r + 5$, so the

    eigenvalues are $r = 1 \pm 2i$. The eigenvector

    corresponding to $1 + 2i$ satisfies $\begin{pmatrix} 2-2i & -2 \\ 4 & -2-2i \end{pmatrix} \begin{pmatrix} \xi_1 \\ \xi_2 \end{pmatrix} = \begin{pmatrix} 0 \\ 0 \end{pmatrix}$,

    or $(2-2i) \xi_1 - 2\xi_2 = 0$. If $\xi_1 = 1$, then $\xi_2 = 1-i$ and

    $\underset{\sim}{\xi}^{(1)} = \begin{pmatrix} 1 \\ 1-i \end{pmatrix}$ and thus one complex-valued solution of the

    D.E. is $\underset{\sim}{x}^{(1)}(t) = \begin{pmatrix} 1 \\ 1-i \end{pmatrix} e^{(1+2i)t}$. To find real-valued solutions (see Eqs.8 and 9) we take the real and

imaginary parts, respectively of $x^{(1)}(t)$. Thus

$$x^{(1)}(t) = \begin{pmatrix} 1 \\ 1-i \end{pmatrix} e^t(\cos 2t + i \sin 2t)$$

$$= e^t \begin{pmatrix} \cos 2t + i \sin 2t \\ \cos 2t + \sin 2t + i(\sin 2t - \cos 2t) \end{pmatrix}$$

$$= e^t \begin{pmatrix} \cos 2t \\ \cos 2t + \sin 2t \end{pmatrix} + ie^t \begin{pmatrix} \sin 2t \\ \sin 2t - \cos 2t \end{pmatrix}$$

Hence the general solution of the D.E. is

$$x = c_1 e^t \begin{pmatrix} \cos 2t \\ \cos 2t + \sin 2t \end{pmatrix} + c_2 e^t \begin{pmatrix} \sin 2t \\ \sin 2t - \cos 2t \end{pmatrix}.$$

7.   The eigenvalues and eigenvectors of the coefficient

matrix satisfy $\begin{pmatrix} 1-r & 0 & 0 \\ 2 & 1-r & -2 \\ 3 & 2 & 1-r \end{pmatrix} \begin{pmatrix} \xi_1 \\ \xi_2 \\ \xi_3 \end{pmatrix} = \begin{pmatrix} 0 \\ 0 \\ 0 \end{pmatrix}$.   The

determinant of coefficients reduces to $(1-r)(r^2 - 2r + 5)$

so the eigenvalues are $r_1 = 1$, $r_2 = 1 + 2i$, and $r_3 = 1 -$

2i.   The eigenvector corresponding to r_1 satisfies

$\begin{pmatrix} 0 & 0 & 0 \\ 2 & 0 & -2 \\ 3 & 2 & 0 \end{pmatrix} \begin{pmatrix} \xi_1 \\ \xi_2 \\ \xi_3 \end{pmatrix} = \begin{pmatrix} 0 \\ 0 \\ 0 \end{pmatrix}$; hence $\xi_1 - \xi_3 = 0$ and $3\xi_1 +$

$2\xi_2 = 0$.   If we let $\xi_2 = -3$ then $\xi_1 = 2$ and $\xi_3 = 2$, so

one solution of the D.E. is $\begin{pmatrix} 2 \\ -3 \\ 2 \end{pmatrix} e^t$.   The eigenvector corresponding to r_2 satisfies $\begin{pmatrix} 2i & 0 & 0 \\ 2 & -2i & -2 \\ 3 & 2 & -2i \end{pmatrix} \begin{pmatrix} \xi_1 \\ \xi_2 \\ \xi_3 \end{pmatrix} = \begin{pmatrix} 0 \\ 0 \\ 0 \end{pmatrix}$.

Hence $\xi_1 = 0$ and $i\xi_2 + \xi_3 = 0$. If we let $\xi_2 = 1$, then ξ_3

$= -i$. Thus a complex-valued solution is $\begin{pmatrix} 0 \\ 1 \\ -i \end{pmatrix} e^t$.

$(\cos 2t + i \sin 2t)$. Taking the real and imaginary parts

we obtain $\begin{pmatrix} 0 \\ \cos 2t \\ \sin 2t \end{pmatrix} e^t$ and $\begin{pmatrix} 0 \\ \sin 2t \\ -\cos 2t \end{pmatrix} e^t$,

respectively. Thus the general solution is

$$\underset{\sim}{x} = c_1 \begin{pmatrix} 2 \\ -3 \\ 2 \end{pmatrix} e^t + c_2 e^t \begin{pmatrix} 0 \\ \cos 2t \\ \sin 2t \end{pmatrix} + c_3 e^t \begin{pmatrix} 0 \\ \sin 2t \\ -\cos 2t \end{pmatrix}.$$

9. The eigenvalues and eigenvectors of the coefficient

matrix satisfy $\begin{pmatrix} 1-r & -5 \\ 1 & -3-r \end{pmatrix} \begin{pmatrix} \xi_1 \\ \xi_2 \end{pmatrix} = \begin{pmatrix} 0 \\ 0 \end{pmatrix}$. The

determinant of coefficients is $r^2 + 2r + 2$ so that the

eigenvalues are $r = -1 \pm i$. The eigenvector

corresponding to $r = -1 + i$ is given by

$\begin{pmatrix} 2-i & -5 \\ 1 & -2-i \end{pmatrix} \begin{pmatrix} \xi_1 \\ \xi_2 \end{pmatrix} = \underset{\sim}{0}$ so that $\xi_1 = (2+i)\xi_2$ and thus

one complex-valued solution is $\underset{\sim}{x}^{(1)}(t) = \begin{pmatrix} 2+i \\ 1 \end{pmatrix} e^{(-1+i)t}$.

Finding the real and complex parts of $x^{(1)}$ leads

to the general solution

$$\underset{\sim}{x} = c_1 e^{-t} \begin{pmatrix} 2 \cos t - \sin t \\ \cos t \end{pmatrix} + c_2 e^{-t} \begin{pmatrix} 2 \sin t + \cos t \\ \sin t \end{pmatrix}.$$

Setting $t = 0$ we find $\underset{\sim}{x}(0) = \begin{pmatrix} 1 \\ 1 \end{pmatrix} = c_1 \begin{pmatrix} 2 \\ 1 \end{pmatrix} + c_2 \begin{pmatrix} 1 \\ 0 \end{pmatrix}$,

which is equivalent to the system $\begin{aligned} 2c_1 + c_2 &= 1 \\ c_1 + 0 &= 1. \end{aligned}$

Thus $c_1 = 1$ and $c_2 = -1$ and

$$\underset{\sim}{x}(t) = e^{-t} \begin{pmatrix} 2 \cos t - \sin t \\ \cos t \end{pmatrix} - e^{-t} \begin{pmatrix} 2 \sin t + \cos t \\ \sin t \end{pmatrix}$$

$$= e^{-t} \begin{pmatrix} \cos t - 3 \sin t \\ \cos t - \sin t \end{pmatrix} .$$

11. If we seek solutions of the form $\underset{\sim}{x} = \underset{\sim}{\xi} t^r$, then r must be

an eigenvalue and $\underset{\sim}{\xi}$ a corresponding eigenvector of the

coefficient matrix. Thus r and $\underset{\sim}{\xi}$ satisfy

$\begin{pmatrix} -1-r & -1 \\ 2 & -1-r \end{pmatrix} \begin{pmatrix} \xi_1 \\ \xi_2 \end{pmatrix} = \begin{pmatrix} 0 \\ 0 \end{pmatrix}$. The determinant of

coefficients is $(-1-r)^2 + 2 = r^2 + 2r + 3$, so the

eigenvalues are $r = -1 \pm \sqrt{2}\, i$. The eigenvector

corresponding to $-1 + \sqrt{2}\, i$ satisfies

$\begin{pmatrix} -\sqrt{2}\, i & -1 \\ 2 & -\sqrt{2}i \end{pmatrix} \begin{pmatrix} \xi_1 \\ \xi_2 \end{pmatrix} = \begin{pmatrix} 0 \\ 0 \end{pmatrix}$ or $\sqrt{2}\, i\, \xi_1 + \xi_2 = 0.$

If we let $\xi_1 = 1$, then $\xi_2 = -\sqrt{2}\, i$, and $\underset{\sim}{\xi}^{(1)} = \begin{pmatrix} 1 \\ -\sqrt{2}\, i \end{pmatrix}$.

Thus a complex-valued solution of the given D.E. is

$\begin{pmatrix} 1 \\ -\sqrt{2}\, i \end{pmatrix} t^{-1+\sqrt{2}i}$. Referring to Section 4.4, if necessary, we have $t^{-1+\sqrt{2}\,i} = t^{-1}[\cos(\sqrt{2}\,\ln t)$

$+ i\,\sin(\sqrt{2}\,\ln t)]$ for $t > 0$. Separating the complex

valued solution into real and imaginary parts, we obtain

the two real-valued solutions $\underset{\sim}{u} = t^{-1}\begin{pmatrix} \cos(\sqrt{2}\,\ln t) \\ 2\,\sin(\sqrt{2}\,\ln t) \end{pmatrix}$

$\underset{\sim}{v} = t^{-1}\begin{pmatrix} \sin(\sqrt{2}\,\ln t) \\ -2\,\cos(\sqrt{2}\,\ln t) \end{pmatrix}$.

Section 7.7, Page 380

1. The eigenvalues and eigenvectors of the given coefficient

matrix satisfy $\begin{pmatrix} 3-r & -4 \\ 1 & -1-r \end{pmatrix}\begin{pmatrix} \xi_1 \\ \xi_2 \end{pmatrix} = \begin{pmatrix} 0 \\ 0 \end{pmatrix}$. The determinant

of coefficients is $(3-r)(-1-r) + 4 = r^2 - 2r + 1 = (r-1)^2$

so $r_1 = 1$ and $r_2 = 1$. The eigenvectors corresponding to

this double eigenvalue satisfy $\begin{pmatrix} 2 & -4 \\ 1 & -2 \end{pmatrix}\begin{pmatrix} \xi_1 \\ \xi_2 \end{pmatrix} = \begin{pmatrix} 0 \\ 0 \end{pmatrix}$,

or $\xi_1 - 2\xi_2 = 0$. Thus the only eigenvectors are

multiples of $\underset{\sim}{\xi}^{(1)} = \begin{pmatrix} 2 \\ 1 \end{pmatrix}$. One solution of the given

D.E. is $\underset{\sim}{x}^{(1)}(t) = \begin{pmatrix} 2 \\ 1 \end{pmatrix} e^t$, but there is no second solution

of this form. To find a second solution we assume that

$\underset{\sim}{x} = \underset{\sim}{\eta}te^t + \underset{\sim}{\zeta}e^t$ and substitute this expression into the

D.E. As in Example 1 we find that $\underset{\sim}{\eta}$ is an eigenvector,

so we choose $\underset{\sim}{\eta} = \begin{pmatrix} 2 \\ 1 \end{pmatrix}$. Then $\underset{\sim}{\zeta}$ must satisfy

$\begin{pmatrix} 2 & -4 \\ 1 & -2 \end{pmatrix}\begin{pmatrix} \zeta_1 \\ \zeta_2 \end{pmatrix} = \begin{pmatrix} 2 \\ 1 \end{pmatrix}$, which verifies Eq.(20). Solving

these equations yields $\zeta_1 - 2\zeta_2 = 1$. If $\zeta_2 = k$, where k is an arbitrary constant, then $\zeta_1 = 1 + 2k$. Hence the

second solution that we obtain is $\underset{\sim}{x}^{(2)}(t) =$

$$\begin{pmatrix} 2 \\ 1 \end{pmatrix} t e^t + \begin{pmatrix} 1 + 2k \\ k \end{pmatrix} e^t = \begin{pmatrix} 2 \\ 1 \end{pmatrix} t e^t + \begin{pmatrix} 1 \\ 0 \end{pmatrix} e^t + k \begin{pmatrix} 2 \\ 1 \end{pmatrix} e^t.$$

The last term is a multiple of the first solution $\underset{\sim}{x}^{(1)}(t)$

and may be neglected, that is, we may set $k = 0$. Thus

$$\underset{\sim}{x}^{(2)}(t) = \begin{pmatrix} 2 \\ 1 \end{pmatrix} t e^t + \begin{pmatrix} 1 \\ 0 \end{pmatrix} e^t \text{ and the general solution is}$$

$$\underset{\sim}{x} = c_1 \underset{\sim}{x}^{(1)}(t) + c_2 \underset{\sim}{x}^{(2)}(t).$$

5.   Substituting $\underset{\sim}{x} = \underset{\sim}{\xi} e^{rt}$ into the given system, we find that

the eigenvalues and eigenvectors satisfy

$$\begin{pmatrix} 1-r & 1 & 1 \\ 2 & 1-r & -1 \\ 0 & -1 & 1-r \end{pmatrix} \begin{pmatrix} \xi_1 \\ \xi_2 \\ \xi_3 \end{pmatrix} = \begin{pmatrix} 0 \\ 0 \\ 0 \end{pmatrix}. \text{ The determinant of}$$

coefficients is $-r^3 + 3r^2 - 4$ and thus $r_1 = -1$, $r_2 = 2$

and $r_3 = 2$. The eigenvector corresponding to r_1

satisfies $\begin{pmatrix} 2 & 1 & 1 \\ 2 & 2 & -1 \\ 0 & -1 & 2 \end{pmatrix} \begin{pmatrix} \xi_1 \\ \xi_2 \\ \xi_3 \end{pmatrix} = \begin{pmatrix} 0 \\ 0 \\ 0 \end{pmatrix}$ which yields $\underset{\sim}{\xi}^{(1)} =$

$$\begin{pmatrix} -3 \\ 4 \\ 2 \end{pmatrix} \text{ and } \underset{\sim}{x}^{(1)} = \begin{pmatrix} -3 \\ 4 \\ 2 \end{pmatrix} e^{-t}. \text{ The eigenvectors}$$

corresponding to the double eigenvalue must satisfy

$$\begin{pmatrix} -1 & 1 & 1 \\ 2 & -1 & -1 \\ 0 & -1 & -1 \end{pmatrix} \begin{pmatrix} \xi_1 \\ \xi_2 \\ \xi_3 \end{pmatrix} = \begin{pmatrix} 0 \\ 0 \\ 0 \end{pmatrix}$$, which yields the single

eigenvector $\xi^{(2)} = \begin{pmatrix} 0 \\ 1 \\ -1 \end{pmatrix}$ and hence $x^{(2)}(t) = \begin{pmatrix} 0 \\ 1 \\ -1 \end{pmatrix} e^{2t}.$

The second solution corresponding to the double

eigenvalue will have the form specified by Eq.(19), which

yields $x^{(3)} = \begin{pmatrix} 0 \\ 1 \\ -1 \end{pmatrix} te^{2t} + \eta e^{2t}.$ Substituting this into

the given system, or using Eq.(20), we find that η

satisfies $\begin{pmatrix} -1 & 1 & 1 \\ 2 & -1 & -1 \\ 0 & -1 & -1 \end{pmatrix} \begin{pmatrix} \eta_1 \\ \eta_2 \\ \eta_3 \end{pmatrix} = \begin{pmatrix} 0 \\ 1 \\ -1 \end{pmatrix}.$ Using row

reduction we find that $\eta_1 = 1$ and $\eta_2 + \eta_3 = 1$, where

either η_2 or η_3 is arbitrary. If we choose $\eta_2 = 0$, then

$\eta = \begin{pmatrix} 1 \\ 0 \\ 1 \end{pmatrix}$ and thus $x^{(3)} = \begin{pmatrix} 0 \\ 1 \\ -1 \end{pmatrix} te^{2t} + \begin{pmatrix} 1 \\ 0 \\ 1 \end{pmatrix} e^{2t}.$ The

general solution is then $x = c_1 x^{(1)} + c_2 x^{(2)} + c_3 x^{(3)}.$

9.   The eigenvalues and eigenvectors of the coefficient

matrix satisfy $\begin{pmatrix} 1-r & 1 & 1 \\ 2 & 1-r & -1 \\ -3 & 2 & 4-r \end{pmatrix} \begin{pmatrix} \xi_1 \\ \xi_2 \\ \xi_3 \end{pmatrix} = \begin{pmatrix} 0 \\ 0 \\ 0 \end{pmatrix}.$ The determinant of coefficients is $8 - 12r + 6r^2 - r^3 =$ $(2-r)^3$, so the eigenvalues are $r_1 = r_2 = r_3 = 2.$ The eigenvectors corresponding to this triple eigenvalue

satisfy $\begin{pmatrix} -1 & 1 & 1 \\ 2 & -1 & -1 \\ -3 & 2 & 2 \end{pmatrix} \begin{pmatrix} \xi_1 \\ \xi_2 \\ \xi_3 \end{pmatrix} = \begin{pmatrix} 0 \\ 0 \\ 0 \end{pmatrix}.$ Using row reduction

we can reduce this to the equivalent system $\xi_1 - \xi_2 - \xi_3 = 0$, $\xi_2 + \xi_3 = 0.$ If we let $\xi_2 = 1$, then $\xi_3 = -1$ and ξ_1 = 0, so the only eigenvectors are multiples of $\xi = \begin{pmatrix} 0 \\ 1 \\ -1 \end{pmatrix}.$

Thus one solution of the given D.E. is $\underset{\sim}{x}^{(1)}(t) = \begin{pmatrix} 0 \\ 1 \\ -1 \end{pmatrix} e^{2t},$ but there are no other linearly independent solutions of this form. We now seek a second solution of the form $\underset{\sim}{x} =$ $\underset{\sim}{\xi} t e^{2t} + \underset{\sim}{\eta} e^{2t}.$ As in the text, $\underset{\sim}{\xi}$ must be an eigenvector

so we choose $\underset{\sim}{\xi} = \begin{pmatrix} 0 \\ 1 \\ -1 \end{pmatrix}.$ Then $\underset{\sim}{\eta}$ must satisfy a system of

the form given by Eq.(20), that is

$\begin{pmatrix} -1 & 1 & 1 \\ 2 & -1 & -1 \\ -3 & 2 & 2 \end{pmatrix} \begin{pmatrix} \eta_1 \\ \eta_2 \\ \eta_3 \end{pmatrix} = \begin{pmatrix} 0 \\ 1 \\ -1 \end{pmatrix}.$ By row reduction this is equivalent to the system $\begin{pmatrix} 1 & -1 & -1 \\ 0 & 1 & 1 \\ 0 & 0 & 0 \end{pmatrix} \begin{pmatrix} \eta_1 \\ \eta_2 \\ \eta_3 \end{pmatrix} = \begin{pmatrix} 0 \\ 1 \\ 0 \end{pmatrix}$. If

we choose $\eta_2 = 0$, then $\eta_3 = 1$ and $\eta_1 = 1$, so $\underset{\sim}{\eta} = \begin{pmatrix} 1 \\ 0 \\ 1 \end{pmatrix}$.

Hence a second solution of the D.E. is $\underset{\sim}{x}^{(2)}(t) =$

$\begin{pmatrix} 0 \\ 1 \\ -1 \end{pmatrix} te^{2t} + \begin{pmatrix} 1 \\ 0 \\ 1 \end{pmatrix} e^{2t}$. Finally, we seek a third

solution of the form $\underset{\sim}{x} = \underset{\sim}{\xi}(t^2/2)e^{2t} + \underset{\sim}{\eta}te^{2t} + \underset{\sim}{\zeta}e^{2t}$, from

Eq.(22), where $\underset{\sim}{\xi}$ and $\underset{\sim}{\eta}$ are as above, and $\underset{\sim}{\zeta}$ satisfies a

system of the form given by Eq.(23), namely

$\begin{pmatrix} -1 & 1 & 1 \\ 2 & -1 & -1 \\ -3 & 2 & 2 \end{pmatrix} \begin{pmatrix} \zeta_1 \\ \zeta_2 \\ \zeta_3 \end{pmatrix} = \begin{pmatrix} 1 \\ 0 \\ 1 \end{pmatrix}$.

By row reduction we find the equivalent system

$\begin{pmatrix} 1 & -1 & -1 \\ 0 & 1 & 1 \\ 0 & 0 & 0 \end{pmatrix} \begin{pmatrix} \zeta_1 \\ \zeta_2 \\ \zeta_3 \end{pmatrix} = \begin{pmatrix} -1 \\ 2 \\ 0 \end{pmatrix}$. If we let $\zeta_2 = 0$, then

$\zeta_3 = 2$ and $\zeta_1 = 1$, so $\underset{\sim}{\zeta} = \begin{pmatrix} 1 \\ 0 \\ 2 \end{pmatrix}$. Since any multiple of a

solution is again a solution it is convenient to multiply

by 2 and write $\underset{\sim}{x}^{(3)}(t) =$

$$\begin{pmatrix} 0 \\ 1 \\ -1 \end{pmatrix} t^2 e^{2t} + 2 \begin{pmatrix} 1 \\ 0 \\ 1 \end{pmatrix} t e^{2t} + 2 \begin{pmatrix} 1 \\ 0 \\ 2 \end{pmatrix} e^{2t}. \quad \text{The general}$$

solution is then $\underset{\sim}{x} = c_1 \underset{\sim}{x}^{(1)}(t) + c_2 \underset{\sim}{x}^{(2)}(t) + c_3 \underset{\sim}{x}^{(3)}(t)$.

11. Assuming $\underset{\sim}{x} = \underset{\sim}{\xi} t^r$ and substituting into the given system,

we find r and $\underset{\sim}{\xi}$ must satisfy $\begin{pmatrix} 1-r & -4 \\ 4 & -7-r \end{pmatrix} \begin{pmatrix} \xi_1 \\ \xi_2 \end{pmatrix} = \begin{pmatrix} 0 \\ 0 \end{pmatrix}$,

which has the double eigenvalue r = -3 and single

eigenvector $\begin{pmatrix} 1 \\ 1 \end{pmatrix}$. Hence one solution of the given D.E.

is $\underset{\sim}{x}^{(1)}(t) = \begin{pmatrix} 1 \\ 1 \end{pmatrix} t^{-3}$. By analogy with the scalar case

considered in Section 4.4 and Example 1 of this section,

we seek a second solution of the form $\underset{\sim}{x} = \underset{\sim}{\eta} t^{-3} \ln t +$

$\underset{\sim}{\zeta} t^{-3}$. Substituting this expression into the D.E. we find

that $\underset{\sim}{\eta}$ and $\underset{\sim}{\zeta}$ satisfy the equations $(\underset{\sim}{A} - 3\underset{\sim}{I})\underset{\sim}{\eta} = \underset{\sim}{0}$ and

$(\underset{\sim}{A} - 3\underset{\sim}{I})\underset{\sim}{\zeta} = \underset{\sim}{\eta}$, where $\underset{\sim}{A} = \begin{pmatrix} 1 & -4 \\ 4 & -7 \end{pmatrix}$ and $\underset{\sim}{I}$ is the identity

matrix. Thus $\underset{\sim}{\eta} = \begin{pmatrix} 1 \\ 1 \end{pmatrix}$, from above, and $\underset{\sim}{\zeta}$ is found to be

$\begin{pmatrix} 0 \\ -1/4 \end{pmatrix}$. Thus a second solution is

$\underset{\sim}{x}^{(2)}(t) = \begin{pmatrix} 1 \\ 1 \end{pmatrix} t^{-3} \ln t + \begin{pmatrix} 0 \\ -1/4 \end{pmatrix} t^{-3}$.

13. All solutions of the given system approach zero as $t \to \infty$

if and only if the eigenvalues of the coefficient matrix

either are real and negative or else are complex with

negative real part. Write down the determinantal

equation satisfied by the eigenvalues and determine when

the eigenvalues are as stated.

Section 7.8, Page 386

Each of the Problems 1 through 10 has been solved in one of
the previous sections. Thus a fundamental matrix for the given
systems can be readily written down. The fundamental matrix
$\underset{\sim}{\Phi}(t)$ satisfying $\underset{\sim}{\Phi}(0) = I$ can then be found, as shown in the
following problems.

4.   From Problem 4 of Section 7.5 we have the two linearly

independent solutions $\underset{\sim}{x}^{(1)}(t) = \begin{pmatrix} 1 \\ -4 \end{pmatrix} e^{-3t}$ and $\underset{\sim}{x}^{(2)}(t) =$

$\begin{pmatrix} 1 \\ 1 \end{pmatrix} e^{2t}$. Hence a fundamental matrix $\underset{\sim}{\Psi}$ is given by $\underset{\sim}{\Psi}(t)$

$= \begin{pmatrix} e^{-3t} & e^{2t} \\ -4e^{-3t} & e^{2t} \end{pmatrix}$. To find the fundamental matrix $\underset{\sim}{\Phi}(t)$

satisfying the I.C. $\underset{\sim}{\Phi}(0) = \underset{\sim}{I}$ we can proceed in either of

two ways. One way is to find $\underset{\sim}{\Psi}(0)$, invert it to obtain

$\underset{\sim}{\Psi}^{-1}(0)$, and then to form the product $\underset{\sim}{\Psi}(t)\underset{\sim}{\Psi}^{-1}(0)$, which

is $\underset{\sim}{\Phi}(t)$. Alternatively, we can find the first column of $\underset{\sim}{\Phi}$

by determining the linear combination $c_1\underset{\sim}{x}^{(1)}(t) +$

$c_2\underset{\sim}{x}^{(2)}(t)$ that satisfies the I.C. $\begin{pmatrix} 1 \\ 0 \end{pmatrix}$. This requires

that $c_1 + c_2 = 1$, $-4c_1 + c_2 = 0$, so we obtain $c_1 = 1/5$

and $c_2 = 4/5$. Thus the first column of $\underset{\sim}{\Phi}(t)$ is

$\begin{pmatrix} (1/5)e^{-3t} + (4/5)e^{2t} \\ -(4/5)e^{-3t} + (4/5)e^{2t} \end{pmatrix}$. Similarly, the second

column of $\underset{\sim}{\Phi}$ is that linear combination of $\underset{\sim}{x}^{(1)}(t)$ and

$x^{(2)}(t)$ that satisfies the I.C. $\begin{pmatrix} 0 \\ 1 \end{pmatrix}$. Thus we must have

$c_1 + c_2 = 0$, $-4c_1 + c_2 = 1$; therefore $c_1 = -1/5$

and $c_2 = 1/5$. Hence the second column of $\Phi(t)$ is

$$\begin{pmatrix} -(1/5)e^{-3t} & + & (1/5)e^{2t} \\ (4/5)e^{-3t} & + & (1/5)e^{2t} \end{pmatrix}.$$

6. Two linearly independent real-valued solutions of the
    given D.E. were found in Problem 2 of Section 7.6. Using
    the result of that problem, we have $\Psi(t) =$

$$\begin{pmatrix} -2e^{-t}\sin 2t & 2e^{-t}\cos 2t \\ e^{-t}\cos 2t & e^{-t}\sin 2t \end{pmatrix}.$$ To find $\Phi(t)$ we

determine the linear combinations of the columns of $\Psi(t)$
that satisfy the I.C. $\begin{pmatrix} 1 \\ 0 \end{pmatrix}$ and $\begin{pmatrix} 0 \\ 1 \end{pmatrix}$, respectively. In
the first case c_1 and c_2 satisfy $0c_1 + 2c_2 = 1$ and $c_1 +$
$0c_2 = 0$. Thus $c_1 = 0$ and $c_2 = 1/2$. In the second case
we have $0c_1 + 2c_2 = 0$ and $c_1 + 0c_2 = 1$, so $c_1 = 1$ and c_2
$= 0$. Using these values of c_1 and c_2 to form the first
and second columns of $\Phi(t)$ respectively, we obtain

$$\Phi(t) = \begin{pmatrix} e^{-t}\cos 2t & -2e^{-t}\sin 2t \\ (1/2)e^{-t}\sin 2t & e^{-t}\cos 2t \end{pmatrix}.$$

8. Two linearly independent solutions of this D.E. were
    found in Problem 1 of Section 7.7. Using that result we

have $\Psi(t) = \begin{pmatrix} 2e^t & e^t + 2te^t \\ e^t & te^t \end{pmatrix}.$ To find $\Phi(t)$ we determine the linear combinations of the columns of $\underset{\sim}{\Psi}(t)$ that satisfy the I.C. $\begin{pmatrix} 1 \\ 0 \end{pmatrix}$ and $\begin{pmatrix} 0 \\ 1 \end{pmatrix}$, respectively. In the first case we have $2c_1 + c_2 = 1$ and $c_1 + 0c_2 = 0$, so $c_1 = 0$ and $c_2 = 1$. In the second case $2c_1 + c_2 = 0$ and $c_1 + 0c_2 = 1$, so $c_1 = 1$ and $c_2 = -2$. Using these values of c_1 and c_2 to form the respective columns of $\underset{\sim}{\Phi}(t)$ we obtain $\underset{\sim}{\Phi}(t) = \begin{pmatrix} e^t + 2te^t & -4te^t \\ te^t & e^t - 2te^t \end{pmatrix}$.

Section 7.9, Page 394

1. From Section 7.5 Problem 3 we have $\underset{\sim}{x}^{(c)} = c_1 \begin{pmatrix} 1 \\ 1 \end{pmatrix} e^t + c_2 \begin{pmatrix} 1 \\ 3 \end{pmatrix} e^{-t}$. Note that $\underset{\sim}{g}(t) = \begin{pmatrix} 1 \\ 0 \end{pmatrix} e^t + \begin{pmatrix} 0 \\ 1 \end{pmatrix} t$ and that $r = 1$ is an eigenvalue of the coefficient matrix. Thus if the method of undetermined coefficients is used, the assumed form is given by Eq.(18).

2. Using methods of previous sections, we find that the eigenvalues are $r_1 = 2$ and $r_2 = -2$, with corresponding eigenvectors $\begin{pmatrix} \sqrt{3} \\ 1 \end{pmatrix}$ and $\begin{pmatrix} 1 \\ -\sqrt{3} \end{pmatrix}$. Thus $\underset{\sim}{x}^{(c)} = c_1 \begin{pmatrix} \sqrt{3} \\ 1 \end{pmatrix} e^{2t} + c_2 \begin{pmatrix} 1 \\ -\sqrt{3} \end{pmatrix} e^{-2t}$. Writing the nonhomogeneous term as $\begin{pmatrix} 1 \\ 0 \end{pmatrix} e^t + \begin{pmatrix} 0 \\ \sqrt{3} \end{pmatrix} e^{-t}$ we see that we can assume $\underset{\sim}{x}^{(p)} = \underset{\sim}{a} e^t + \underset{\sim}{b} e^{-t}$. Substituting this in the D.E., we obtain $\underset{\sim}{a} e^t - \underset{\sim}{b} e^{-t} = \underset{\sim}{A} \underset{\sim}{a} e^t + \underset{\sim}{A} \underset{\sim}{b} e^{-t} + \begin{pmatrix} 1 \\ 0 \end{pmatrix} e^t + \begin{pmatrix} 0 \\ \sqrt{3} \end{pmatrix} e^{-t}$, where $\underset{\sim}{A}$ is

the given coefficient matrix. All the terms involving e^t must add to zero and thus we have $\underset{\sim}{A} \underset{\sim}{a} - \underset{\sim}{a} + \begin{pmatrix} 1 \\ 0 \end{pmatrix} = \begin{pmatrix} 0 \\ 0 \end{pmatrix}$.

This is equivalent to the system $\sqrt{3}\, a_2 = -1$ and $\sqrt{3}\, a_1 - 2a_2 = 0$, or $a_1 = -2/3$ and $a_2 = -1/\sqrt{3}$. Likewise the terms involving e^{-t} must add to zero, which yields $\underset{\sim}{A} \underset{\sim}{b} + \underset{\sim}{b} + \begin{pmatrix} 0 \\ \sqrt{3} \end{pmatrix} = \begin{pmatrix} 0 \\ 0 \end{pmatrix}$. The solution of this sytem is $b_1 = -1$

and $b_2 = 2/\sqrt{3}$. Substituting these values for $\underset{\sim}{a}$ and $\underset{\sim}{b}$ into $\underset{\sim}{x}^{(p)}$ and adding $\underset{\sim}{x}^{(p)}$ to $\underset{\sim}{x}^{(c)}$ yields the desired solution.

3. The method of undetermined coefficients is not straight forward since the assumed form of $\underset{\sim}{x}^{(p)} = \underset{\sim}{a} \cos t + \underset{\sim}{b} \sin t$ leads to singular equations for $\underset{\sim}{a}$ and $\underset{\sim}{b}$. From Problem 3 of Section 7.6 we find that a fundamental matrix is $\underset{\sim}{\Psi}(t) = \begin{pmatrix} 5 \cos t & 5 \sin t \\ 2 \cos t + \sin t & -\cos t + 2 \sin t \end{pmatrix}$.

The inverse matrix is found (using steps similar to those illustrated in Example 2 of Section 7.2) to be $\underset{\sim}{\Psi}^{-1}(t) = \begin{pmatrix} \dfrac{\cos t - 2 \sin t}{5} & \sin t \\ \dfrac{2 \cos t + \sin t}{5} & -\cos t \end{pmatrix}$. Thus we may use the

method of variation of parameters where $\underset{\sim}{x} = \underset{\sim}{\Psi}(t)\, \underset{\sim}{u}(t)$ and $\underset{\sim}{u}(t)$ is given by $\underset{\sim}{u}'(t) = \underset{\sim}{\Psi}^{-1}(t)\underset{\sim}{g}(t)$ from Eq.(27). For this problem $\underset{\sim}{g}(t) = \begin{pmatrix} -\cos t \\ \sin t \end{pmatrix}$ and thus

$$\underset{\sim}{u}'(t) = \begin{pmatrix} \dfrac{\cos t - 2 \sin t}{5} & \sin t \\[3mm] \dfrac{2 \cos t + \sin t}{5} & -\cos t \end{pmatrix} \begin{pmatrix} -\cos t \\[3mm] \sin t \end{pmatrix}$$

$$= \frac{1}{5} \begin{pmatrix} 2 - 3 \cos 2t + \sin 2t \\ -1 - \cos 2t - 3 \sin 2t \end{pmatrix} ,$$

after multiplying and using appropriate trigonometric identities. Integration and multiplication by $\underset{\sim}{\Psi}$ yields the desired solution.

4.   In this problem we use the method illustrated in Example 1. From Problem 4 of Section 7.5 we have the transformation matrix

$$\underset{\sim}{T} = \begin{pmatrix} 1 & 1 \\ -4 & 1 \end{pmatrix} . \text{ Inverting } \underset{\sim}{T} \text{ we find that } \underset{\sim}{T}^{-1} = \frac{1}{5} \begin{pmatrix} 1 & -1 \\ 4 & 1 \end{pmatrix} .$$

If we let $\underset{\sim}{x} = \underset{\sim}{T}\underset{\sim}{y}$ and substitute into the D.E., we obtain

$$\underset{\sim}{y}' = \frac{1}{5} \begin{pmatrix} 1 & -1 \\ 4 & 1 \end{pmatrix} \begin{pmatrix} 1 & 1 \\ 4 & -2 \end{pmatrix} \begin{pmatrix} 1 & 1 \\ -4 & 1 \end{pmatrix} \underset{\sim}{y} + \frac{1}{5} \begin{pmatrix} 1 & -1 \\ 4 & 1 \end{pmatrix} \begin{pmatrix} e^{-2t} \\ -2e^{t} \end{pmatrix}$$

$$= \begin{pmatrix} -3 & 0 \\ 0 & 2 \end{pmatrix} \underset{\sim}{y} + \frac{1}{5} \begin{pmatrix} e^{-2t} + 2e^{t} \\ 4e^{-2t} - 2e^{t} \end{pmatrix} .$$

This corresponds to the two scalar equations

$$y_1' + 3y_1 = (1/5)e^{-2t} + (2/5)e^{t} ,$$

$$y_2' - 2y_2 = (4/5)e^{-2t} - (2/5)e^{t} ,$$

which may be solved by the methods of Section 2.1. For the first equation the integrating factor is e^{3t} and we obtain $(e^{3t}y_1)' = (1/5)e^{t} + (2/5)e^{4t}$, so $e^{3t}y_1 = (1/5)e^{t} + (1/10)e^{4t} + c_1$. For the second equation the integrating factor is e^{-2t}, so $(e^{-2t}y_2)' = (4/5)e^{-4t} -$ $(2/5)e^{-t}$. Hence $e^{-2t}y_2 = - (1/5)e^{-4t} + (2/5)e^{-t} + c_2$.

Thus $y = \begin{pmatrix} 1/5 \\ -1/5 \end{pmatrix} e^{-2t} + \begin{pmatrix} 1/10 \\ 2/5 \end{pmatrix} e^{t} + \begin{pmatrix} c_1 e^{-3t} \\ c_2 e^{2t} \end{pmatrix}$.

Finally, multiplying by T, we obtain

$$x = Ty = \begin{pmatrix} 0 \\ -1 \end{pmatrix} e^{-2t} + \begin{pmatrix} 1/2 \\ 0 \end{pmatrix} e^{t} + c_1 \begin{pmatrix} 1 \\ -4 \end{pmatrix} e^{-3t} + c_2 \begin{pmatrix} 1 \\ 1 \end{pmatrix} e^{2t}.$$

The last two terms are the general solution of the corresponding homogeneous system, while the first two terms are a particular solution of the nonhomogeneous system.

12. Since the coefficient matrix is the same as that of Problem 3, use the same procedure as done in that problem, including the Ψ^{-1} found there. In the interval $\pi/2 < t < \pi$ note that $\sin t > 0$ and $\cos t < 0$; hence $|\sin t| = \sin t$, but $|\cos t| = - \cos t$.

14. To verify that the given vector is the general solution of the corresponding system, it is sufficient to substitute it into the D.E. Note also that the two terms in $x^{(c)}$ are linearly independent. If we seek a solution of the form $x = \Psi(t)u(t)$ then we find that the equation corresponding to Eq.(26) is $t\Psi(t)u'(t) = g(t)$, where

$$\Psi(t) = \begin{pmatrix} t & 1/t \\ t & 3/t \end{pmatrix} \quad \text{and} \quad g(t) = \begin{pmatrix} 1-t^2 \\ 2t \end{pmatrix}. \quad \text{Thus } u' =$$

$(1/t)\Psi^{-1}(t)g(t)$. Using row operations on Ψ and I, we

find that $\Psi^{-1} = \begin{pmatrix} 3/2t & -1/2t \\ -t/2 & t/2 \end{pmatrix}$ and thus $u' =$

$\begin{pmatrix} 3/2t^2 - 3/2 - 1/t \\ -1/2 + t^2/2 + t \end{pmatrix}$. Integration and multiplication

by $\Psi(t)$ yields the desired solution.

CHAPTER 8

Section 8.2, Page 404

2a. The Euler formula is $y_{n+1} = y_n + h(2y_n - x_n + 1/2)$ for n
= 0,1,2,3 and with $x_0 = 0$ and $y_0 = 1$. Thus $y_1 = y_0 + .1(2y_0 - x_0 + 1/2) = 1.25$ and $y_2 = 1.25 + .1[2(1.25) - (.1) + 1/2] = 1.54$

2b. Use the same formula as in Problem 2a, except now h = .05 and n = 0,1...7.

2c. $y' = 1/2 - x + 2y$ is a first order linear D.E. Rewrite the equation in the form $y' - 2y = 1/2 - x$ and multiply both sides by the integrating factor e^{-2x} to obtain $(e^{-2x}y)' = (1/2 - x)e^{-2x}$. Integrating the right side by parts and multiplying by e^{2x} we obtain $y = c\ e^{-2x} + x/2$. The I.C. $y(0) = 1 \rightarrow c = 1$ and hence the solution of the I.V.P. is $y = \phi(x) = e^{2x} + x/2$. Thus $\phi(0.1) = 1.271$, $\phi(0.2) = 1.592$, $\phi(0.3) = 1.972$, and $\phi(0.4) = 2.426$.

5. The Euler formula is $y_{n+1} = y_n + h\sqrt{x_n + y_n}$ for n = 0,1,2... with $x_0 = 1$ and $y_0 = 3$.

7. Use the hint along with the values of y that you calculate for x between 1.6 and 1.8. Note that the slope (y') is positive for values of y < 1.155.

9. If $y' = 1 - x + 4y$, then $y'' = -1 + 4y' = -1 + 4(1 - x + 4y) = 3 - 4x + 16y$. In Eq.(9) we let y_n, y_n' and y_n''

denote the approximate values of $\phi(x_n)$, $\phi'(x_n)$, and $\phi''(x_n)$, respectively. Keeping the first three terms in the Taylor series we have $y_{n+1} = y_n + y_n'h + y_n''h^2/2 = y_n + (1 - x_n + 4y_n)h + (3 - 4x_n + 16y_n)h^2/2$ for $n = 0,1$ with $x_0 = 0$ and $y_0 = 1$.

11b. Using Eq.(3) we have $y_{n+1} = y_n + h(2 y_n - 1) = (1 + 2h)y_n - h$. Setting $n + 1 = k$ (and hence $n = k - 1$) this becomes $y_k = (1 + 2h)y_{k-1} - h$, for $k = 1,2,\ldots$. Since $y_0 = 1$, we have $y_1 = 1 + 2h - h = 1 + h = (1 + 2h)/2 + 1/2$, and hence $y_2 = (1 + 2h)y_1 - h = (1 + 2h)^2/2 + (1 + 2h)/2 - h = (1 + 2h)^2/2 + 1/2$; $y_3 = (1 + 2h)y_2 - h = (1 + 2h)^3/2 + (1 + 2h)/2 - h = (1 + 2h)^3/2 + 1/2$. Continuing in this fashion (or using induction) we obtain $y_k = (1 + 2h)^k/2 + 1/2$. For fixed $x > 0$ choose $h = x/k$. Then substitute for h in the last formula to obtain $y_k = (1 + 2x/k)^k/2 + 1/2$. Letting $k \to \infty$ we find $y(x) = y_k \to e^{2x}/2 + 1/2$, which is the exact solution. (See hint for Problem 10d.)

12a. $\phi_{n+1}(x) = 1 + \int_0^x [2\phi_n(t) - 1]dt$, $\phi_0(x) = 1$

$\phi_1(x) = 1 + \int_0^x (2-1)dt = 1 + x$

$\phi_2(x) = 1 + \int_0^x [2(1+t) - 1]dt = 1 + x + x^2$

$\phi_3(x) = 1 + \int_0^x [2(1+t+t^2) - 1]dt = 1 + x + x^2 + \frac{2}{3}x^3$.

Section 8.3, Page 410

1. If $y = \phi(x)$ is the exact solution of the I.V.P., then

    $\phi'(x) = 2\phi(x) - 1$ and $\phi''(x) = 2\phi'(x) = 4\phi(x) - 2$. From

    Eq.(10), $e_{n+1} = [2\phi(\bar{x}_n) - 1]h^2$, $x_n < \bar{x}_n < x_n + h$. Thus

    $|e_{n+1}| \leq [1 + 2 \max_{0 < x < 1} |\phi(x)|]h^2$. Since the exact solution is

    $y = \phi(x) = [1 + \exp(2x)]/2$, $e_{n+1} = h^2 \exp(2\bar{x}_n)$.

    Therefore $|e_1| \leq (0.1)^2 \exp(0.2) = 0.012$ and $|e_4| \leq$

    $(0.1)^2 \exp(0.8) = 0.022$, since the maximum value of

    $\exp(2\bar{x}_n)$ occurs at $x = .1$ and $x = .4$ respectively.

4. The local formula error is $e_{n+1} = \phi''(\bar{x}_n)h^2/2$. For this

    problem $\phi'(x) = 5x - 3\phi^{1/2}(x)$ and thus $\phi''(x) = 5 -$

    $(3/2)\phi^{-1/2}\phi' = 19/2 - (15/2)x\phi^{-1/2}$. Substituting this

    last expression into e_{n+1} yields the desired answer.

7d. $e_{n+1} = -(5\pi/2)\sin(5\pi\bar{x}_n)h^2$.

8a. From Eq.(3) we have $E_n = \phi(x_n) - y_n$. Using this in

    Eq.(9) we obtain $E_{n+1} = E_n + h\{f[x_n, \phi(x_n)] - f(x_n,y_n)\}$

    $+ \phi''(\bar{x}_n)h^2/2$. Using the given inequality involving L we

    have $|f[x_n, \phi(x_n)] - f(x_n,y_n)| \leq L|\phi(x_n) - y_n| = L|E_n|$ and

    thus $|E_{n+1}| \leq |E_n| + hL|E_n| + \max_{x_0 \leq x \leq x_n} |\phi''(x)| \ h^2/2 = \alpha|E_n| +$

    βh^2.

8b. Since $\alpha = 1 + hL$, $\alpha - 1 = hL$. Hence $\beta h^2(\alpha^n - 1)/(\alpha - 1) =$

    $\beta h^2[(1+hL)^n - 1]/hL = \beta h[(1+hL)^n - 1]/L$.

8c. $(1+hL)^n \leq \exp(nhL)$ follows from the observation that

    $\exp(nhL) = (\exp hL)^n = (1 + hL + h^2L^2/2! + ...)^n$. Noting that $nh = x_n - x_0$, the rest follows from Eq.(ii).

10b. Using a step size of .2 we estimate $\phi(.2) = 1.2$ in one

step. Using a step size of .1 we estimate $\phi(.2) = 1.22$

in two steps. An improved estimate of $\phi(.2)$ is now given

by Eq.(i) of Problem 9, which yields $\phi(.2) = 1.22 +$

$(1.22 - 1.2) = 1.24$.

12. Using the result of Problem 9 and the entries from Table

8.1 we have

    a. $\phi(1) \simeq 45.588400 + 11.176910 = 56.765310$

    b. $\phi(1) \simeq 53.807866 + 8.219466 = 62.027332$

Although the proof suggested in Problem 11 is not

applicable in this case, it is possible to show that the

result is still valid using more advanced analysis.

Thus, since the result of part (a) gives an estimate for

$\phi(1)$ with an error proportional to h^2 with a step size of

.05 and the result of part (b) gives an estimate for $\phi(1)$

with an error proportional to h^2 with a step size of .025

we may obtain a new estimate for $\phi(1)$ with an error

proportional to h^3 with a step size of .025 as follows:

$\phi(1) \simeq 62.027332 + (62.027332 - 56.765310)/(2^2-1) =$

$= 63.781339$.

Section 8.4, Page 413

1. The improved Euler formula is $y_{n+1} = y_n + [y_n' + f(x_n + h,$

$y_n + hy_n')]h/2$ where $y' = f(x,y) = 2y - 1$. Hence $y_n' = 2y_n$

$- 1$, and $f(x_n + h, y_n + hy_n') = 2(y_n + hy_n') - 1$. Thus we

obtain $y_{n+1} = y_n + [y_n' + 2(y_n + hy_n') - 1]h/2$. If

desired, we can substitute for y_n' in terms of y_n and

obtain $y_{n+1} = y_n + h(1+h)(2y_n - 1)$, $n = 0,1,2,3$ with

$y_0 = 1$. In this case the formula for y_{n+1} was made

simpler by substituting for y_n'; but this may not always

be true. Thus $y_1 = 1 + .1(1.1)(1) = 1.11$ and $y_2 = 1.11 +$

$.1(1.1)(1.22) = 1.244$.

5.   The improved Euler formula is $y_{n+1} = y_n + [y_n' + f(x_n + h,$

$y_n + hy_n')]h/2$ where $y' = f(x,y) = \sqrt{x+y}$. Hence $y_n' =$

$\sqrt{x_n+y_n}$ and $f(x_n+h, y_n + hy_n') = \sqrt{(x_n+h)+(y_n+h\sqrt{x_n+y_n})}$.

Thus we obtain

$y_1 = 3 + \dfrac{\sqrt{1+3} + \sqrt{(1.1)+(3+.1\sqrt{1+3})}}{2} (.1) = 3.204$ and

$y_2 = 3.204 +$

$\dfrac{\sqrt{1.1+3.204} + \sqrt{(1.2)+(3.204 + .1\sqrt{1.1+3.204})}}{2} (.1) = 3.415$.

7a.   Since $\phi(x_n + h) = \phi(x_{n+1})$ we have, using Eq.(5) and the

given equation, $e_{n+1} = \phi(x_{n+1}) - y_{n+1} + [\phi(x_n)-y_n] +$

$[\phi'(x_n) - \dfrac{y_n' + f(x_n+h, y_n+hy_n')}{2}]h + \phi''(x_n)h^2/2! +$

$\phi'''(\bar{x}_n)h^3/3!$. Since $y_n = \phi(x_n)$ and $y_n' = \phi'(x_n) =$

$f(x_n,y_n)$ this reduces to $e_{n+1} + \phi''(x_n)h^2/2! - \{f[x_n+h, y_n$

$+ hf(x_n,y_n)] - f(x_n,y_n)\}h/2! + \phi'''(\bar{x}_n)h^3/3!$, which can be

written in the form of Eq.(i).

7b. First observe that $y' = f(x,y)$ and $y'' = f_x(x,y) +$

$f_y(x,y)y'$. Hence $\phi''(x_n) = f_x(x_n,y_n) + f_y(x_n,y_n)f(x_n,y_n)$.

Using the given Taylor series, with $a = x_n$, $h = h$, $b = y_n$

and $k = hf(x_n,y_n)$ we have

$$f[x_n+h,y_n+hf(x_n,y_n)] = f(x_n,y_n)+f_x(x_n,y_n)h+f_y(x_n,y_n)hf(x_n,y_n)$$

$$+ [f_{xx}(\xi,\eta)h^2+2f_{xy}(\xi,\eta)h^2f(x_n,y_n)+f_{yy}(\xi,\eta)h^2f^2(x_n,y_n)]/2!$$

where $x_n < \xi < x_n + h$ and $|\eta-y_n| < h|f(x_n,y_n)|$.

Substituting this in Eq.(i) and using the earlier

expression for $\phi''(x_n)$ we find that the first term on the

right side of Eq.(i) reduces to $- [f_{xx}(\xi,\eta) +$

$2f_{xy}(\xi,\eta)f(x_n,y_n) + f_{yy}(\xi,\eta)f^2(x_n,y_n)]h^3/4$, which is

proportional to h^3 plus, possibly, higher order terms.

The reason that there may be higher order terms is

because ξ and η will, in general, depend upon h.

8.   Since $\phi(x) = [4x - 3 + 19\exp(4x)]/16$ we have $\phi'''(x) =$

$76\exp(4x)$ and thus from Problem7c we find $e_{n+1} =$

$38[\exp(4\bar{x}_n)]h^3/3$. Thus $|e_{n+1}| \leq (38h^3/3)\exp(4) = 691.6h^3$

on $0 \leq x \leq 1$. $|e_1| = |\phi(x_1) - y_1| \leq (0.038/3) \exp(0.4) =$

0.0189, which is approximately 1/10 of the error

indicated in Eq.(10) of the previous section.

11c. The modified Euler formula is $y_{n+1} = y_n + hf[x_n+h/2, y_n +$

$(h/2)f(x_n,y_n)]$ where $f(x,y) = x^2 + y^2$. Since $x_0 = 0$, y_0

$= 1$ and $h = .1$, we have

$y_1 = 1 + .1[(.05)^2 + (1+.05)^2] = 1.1105$ and

$y_2 = 1.1105 + .1 \{(.15)^2 + [1.1105+.05(.1^2+1.1105^2)]^2\} = $

1.2503.

Section 8.5, Page 419

1. If $y' = 2y - 1$ then $y'' = 2y' = 2(2y-1)$. Hence, the

    three-term Taylor series formula is $y_{n+1} = y_n + y_n'h +$

    $y_n''h^2/2 = y_n + y_n'h + y_n'h^2$ with $y_n' = 2y_n - 1$. If desired,

    we can substitute for y_n' and obtain $y_{n+1} = y_n +$

    $h(1+h)(2y_n-1)$, $n = 0,1,2$ with $y_0 = 1$. Thus $y_1 = 1 +$

    $.1(1.1)(1) = 1.11$ and $y_2 = 1.11 + .1(1.1)(1.22) = 1.244$.

    Note that this is the same formula as was obtained using

    the improved Euler method [Problem 1 of Section 8.4],

    which is true whenever f is linear in x and y, as shown

    in Problem 7.

4. Since $y' = 5x - 3\sqrt{y}$, we have $y'' = 5 - (3/2)y'/\sqrt{y} = (19 -$

    $15xy^{-1/2})/2$. Hence the three-term Taylor series formula

    is $y_{n+1} = y_n + h(5x_n-3y_n^{1/2}) + h^2(19-15x_ny_n^{-1/2})/4$. Thus

    $y_1 = 2 + .1(-3\sqrt{2}) + .01(19)/4 = 1.6232$ and $y_2 = 1.6232 +$

    $.1(.5-3\sqrt{1.6232}) + .01[19-15(.1)/\sqrt{1.6232}]/4 = 1.3355$.

9. Since $\phi(x_n+h) = \phi(x_{n+1})$, we have, utilizing Eq.(2) and

    Eq.(6), $e_{n+1} = \phi(x_{n+1}) - y_{n+1} = \phi(x_n) + \phi'(x_n)h +$

    $\phi''(x_n)h^2/2! + \phi'''(\bar{x}_n)h^3/3! - y_n - hy_n' - h^2y_n''/2 =$

    $\phi'''(\bar{x}_n)h^3/3!$ where $x_n \leq \bar{x}_n \leq x_n + h$. Note that we have

    made use of the fact that $y_n = \phi(x_n)$, which means that y_n'

$$= f(x_n, y_n) = f[x_n, \phi(x_n)] = \phi'(x_n) \text{ and from Eq.(8) } y_n'' =$$

$$f_x(x_n, y_n) + f_y(x_n, y_n)y_n' = f_x[x_n, \phi(x_n)] + f_y[x_n, \phi(x_n)] \cdot$$

$$\phi'(x_n) = \phi''(x_n).$$

10. The four-term Taylor series formula for $y' = f(x,y)$ is

$$y_{n+1} = y_n + y_n'h + y_n''h^2/2! + y_n'''h^3/3!, \text{where } y_n' = f(x_n, y_n),$$

$y_n'' = f_x(x_n, y_n) + f_y(x_n, y_n)y_n'$, and $y_n''' = f_{xx}(x_n, y_n) +$

$f_{xy}(x_n, y_n)y_n' + [f_{yx}(x_n, y_n) + f_{yy}(x_n, y_n)y_n']y_n' +$

$f_y(x_n, y_n)y_n'' = f_{xx}(x_n, y_n) + 2f_{xy}(x_n, y_n)y_n' +$

$f_{yy}(x_n, y_n)(y_n')^2 + f_y(x_n, y_n)y_n''$. Using the theory of

Taylor series with a remainder, we find the local formula

error will be $\phi''''(\bar{x}_n)h^4/4!$ where $x_n < \bar{x}_n < x_{n+1}$ and ϕ is

the exact solution of the I.V.P.

Section 8.6, Page 423

The accuracy of the Runge-Kutta method can be seen by
reworking any of the Problems 1 through 6 with a step size of
$h = .1$. In most cases it is found that $h = .2$ gives at least
four place accuracy for these problems.

2. The Runge-Kutta formula is $y_{n+1} = y_n + h(k_{n1} + 2k_{n2} +$

$2k_{n3} + k_{n4})/6$ where k_{n1}, k_{n2} etc. are given by

Eqs.(4). Thus for $f(x,y) = 1/2 - x + 2y$, $(x_0, y_0) =$

$(0,1)$ and $h = .2$ we have

$$k_{01} = f(x_0, y_0) = 1/2 + 2 = 2.5$$

$$k_{02} = f(x_0 + .1, y_0 + .25) = 1/2 - .1 + 2.5 = 2.9$$

$$k_{03} = f(x_0 + .1, y_0 + .29) = 1/2 - .1 + 2.58 = 2.98$$

$$k_{04} = f(x_0 + .2, y_0 + .596) = 1/2 - .2 + 3.192 = 3.492$$

and hence y_1 = 1 + .2(2.5 + 5.8 + 5.96 + 3.492)/6 =

1.59173. To calculate y_2, we take x_1 = .2 and y_1 =

1.59173 and thus

k_{11} = $f(x_1, y_1)$ = 1/2 - .2 + 3.18346 = 3.48346

k_{12} = $f(x_1 + .1, y_1 + .34835)$ = 1/2 - .3 + 3.88016 = 4.08016

k_{13} = $f(x_1 + .1, y_1 + .40802)$ = 1/2 - .3 + 3.99950 = 4.19950

k_{14} = $f(x_1 + .2, y_1 + .83990)$ = 1/2 - .4 + 4.86326 = 4.96326

and hence y_2 = 1.59173 + .2(3.48346 + 8.16032 + 8.39900 +

4.96326)/6 = 2.42526.

5.   The Runge-Kutta formula is y_{n+1} = y_n + $h(k_{n1}$ + $2k_{n2}$ +

$2k_{n3}$ + $k_{n4})/6$ where k_{n1} = $f(x_n, y_n)$ = $\sqrt{x_n + y_n}$, k_{n2} =

$f(x_n + h/2, y_n + h k_{n1}/2)$ = $\sqrt{(x_n + h/2) + (y_n + h k_{n1}/2)}$, k_{n3} =

$f(x_n + h/2, y_n + h k_{n2}/2)$ = $\sqrt{(x_n + h/2) + (y_n + h k_{n2}/2)}$, and

k_{n4} = $f(x_n + h, y_n + h k_{n3})$ = $\sqrt{(x_n + h) + (y_n + h k_{n3})}$. A

sequence of calculations to obtain y_{n+1} would be $(x_n, y_n) \rightarrow$

$k_{n1} \rightarrow k_{n2} \rightarrow k_{n3} \rightarrow k_{n4} \rightarrow y_{n+1}$. In this case we have h =

.2 and for n = 0, x_0 = 1 and y_0 = 3. Thus k_{01} = 2, k_{02} =

2.07364, k_{03} = 2.07542, and k_{04} = 2.14827 and hence y_1 =

3.41488.

8a. The difference between formulas (i) and (ii) is hy_n' +

$h^2 y_n''/2 - h\{af(x_n, y_n) + bf[x_n + \alpha h, y_n + \beta h f(x_n, y_n)]\}$.

Since y' = $f(x_n, y_n)$ we have y_n'' = $f_x(x_n, y_n)$ +

$f_y(x_n, y_n) f(x_n, y_n)$ and from Problem 7 of Section 8.4 we

find (note that h = αh and k = $\beta h f$) $f[x_n + \alpha h, y_n +$

$\beta h f(x_n, y_n)] = f(x_n, y_n) + \alpha h f_x(x_n, y_n) +$

$\beta h f(x_n, y_n) f_y(x_n, y_n)$ + terms involving h^2. Using these

expressions we find that if $a + b = 1$, $b\alpha = 1/2$ and $b\beta = 1/2$, then the difference reduces to h times the terms

involving h^2 and hence the difference is proportional to

h^3.

9. We require that $Ax^2 + Bx + C$ equal $f(x)$ at $x = 0$, $x = h/2$, and $x = h$. Hence $C = f(0)$, $Ah^2/4 + Bh/2 + C = f(h/2)$, and $Ah^2 + Bh + C = f(h)$. The solution of this

system is $A = 2[f(0) - 2f(h/2) + f(h)]/h^2$, $B = [- 3f(0) + 4f(h/2) - f(h)]/h$, and $C = f(0)$. Hence $\int_0^h f(x)dx \approx$

$\int_0^h (Ax^2 + Bx + C)dx = Ah^3/3 + Bh^2/2 + Ch = (h/6) [4f(0) - 8f(h/2) + 4f(h) - 9f(0) + 12f(h/2) - 3f(h) + 6f(0)] = h[f(0) + 4f(h/2) + f(h)]/6$.

Section 8.7, Page 430

2a. If $0 \le x \le 1$ then we know $0 \le x^2 \le 1$ and hence $e^y \le x^2 + e^y \le 1 + e^y$. Since each of these terms represents a

slope, we may conclude that the solution of Eq.(i) is

bounded above by the solution of Eq.(iii) and is bounded

below by the solution of Eq.(iv).

2b. $\phi_1(x)$ and $\phi_2(x)$ can each be found by separation of

variables. For $\phi_1(x)$ we have $\dfrac{1}{1+e^y}$ dy = dx, or $\dfrac{e^{-y}}{e^{-y}+1}$ dy

= dx. Integrating both sides yields $-\ln(e^{-y}+1) = x + c$.

Solving for y we find $y = \ln[1/(c_1 e^{-x}-1)]$. Setting

$x = 0$ and $y = 0$, we obtain $c_1 = 2$ and thus $\phi_1(x) =$

$\ln[e^x/(2-e^x)]$. As $x \rightarrow \ln 2$, we see that $\phi_1(x) \rightarrow \infty$. A

similar analysis shows that $\phi_2(x) = \ln[1/(c_2-x)]$, where

$c_2 = 1$ when the I.C. are used. Thus $\phi_2(x) \rightarrow \infty$ as $x \rightarrow 1$

and thus we conclude that $\phi(x) \rightarrow \infty$ for some x such that

$\ln 2 \leq x \leq 1$.

2c.   Utilize the solutions found in Part b.

3a.   The general solution of the D.E. is $y(x) = x + ce^{\lambda x}$,

where $y(0) = 0 \rightarrow c = 0$ and thus $y(x) = x$, which is

independent of x.

3c.   Your result in part b will depend upon the particular

computer hardware and software that you use. If there is

sufficient accuracy, you will obtain the solution $y = x$

for x on $0 \leq x \leq 1$ for each value of λ that is given,

since there is no discretization error. If there is not

sufficient accuracy, then round-off error will affect

your calculations. For the larger values of λ, the

numerical solution will quickly diverge from the exact

solution, $y = x$, to the general solution $y = x + ce^{\lambda x}$,

where the value of c depends upon the round-off error.

If the latter case does not occur, you may simulate it by

computing the numerical solution to the I.V.P. $y' - \lambda y =$

$1 - \lambda x$, $y(.1) = .10000001$. Here we have assumed that the numerical solution is exact up to the point x = .09

[i.e. y(.09) = .09] and that at x = .1 round-off error

has occurred as indicated by the slight error in the I.C.

It has also been found that a larger step size (h = .05

or h = .1) may also lead to round-off error.

Section 8.8, Page 437

2.   The predictor formula is $y_{n+1} = y_n + (h/24)(55y_n' - 59y_{n-1}'$

$+ 37y_{n-2}' - 9y_{n-3}')$ and the corrector formula is $y_{n+1} = y_n$

$+ (h/24)(9y_{n+1}' + 19y_n' - 5y_{n-1}' + y_{n-2}')$, where $y_n' = 1/2 - x_n$

$+ 2y_n$ in each case. No effort is saved by expressing

y_{n+1} in terms of y_{n-3}, y_{n-2}, y_{n-1} and y_n in these

formulas. Thus given (x_0, y_0), (x_1, y_1), (x_2, y_2) and

$x_3, y_3)$ we may calculate $y_0' = 2.5$, $y_1' = 2.9428$, $y_2' =$

3.48364 and $y_3' = 4.14422$. Thus $y_{4p} = 1.97211 +$

$(.1/24)[55(4.14422) - 59(3.48364) + 37(2.9428) - 9(2.5)]$

$= 2.42536$. This value of y_4 is now used to calculate y_4'

$= 4.95073$ and thus $y_{4c} = 1.97211 + (.1/24)[9(4.95073) +$

$19(4.14422) - 5(3.48364) + 2.9428] = 2.42553$. If values

of y_5, y_6 etc. were desired, the above process would be

repeated, using the value of y_{4c} to find an improved y_4'.

5.   With y' = ax + b we require that $y_{n-1}' = y'(x_{n-1}) = ax_{n-1}$

$+ b = a(x_n - h) + b$ and $y_n' = y'(x_n) = ax_n + b$. Solving the

linear system for a and b yields a = $(y_n' - y_{n-1}')/h$, b =

$[x_n y_{n-1}' - (x_n - h)y']/h$. Next

$$\phi(x_n+1) - \phi(x_{n-1}) = \int_{x_{n-1}}^{x_{n+1}} \phi'(x)dx \simeq \int_{x_n-h}^{x_n+h} (ax+b)dx = 2ahx_n + 2bh$$

$$= 2(y_n' - y_{n-1}')x_n + 2[x_n y_{n-1}' - (x_n-h)y_n'] = 2hy_n'.$$

Setting $\phi(x_n) = y_n$, it follows that $y_{n+1} - y_{n-1} = 2hy_n'$.

6.   We have, using Taylor series:

$$\phi(x_n+h) = \phi(x_n) + h\phi'(x_n) + \phi''(x_n)h^2/2 + \phi'''(\alpha_n)h^3/6$$

and

$$\phi(x_n-h) = \phi(x_n) - h\phi'(x_n) + \phi''(x_n)h^2/2 - \phi'''(\beta_n)h^3/6,$$

where $x_n < \alpha_n < x_n + h$ and $x_n - h < \beta_n < x_n$. Subtracting

the second equation from the first, we obtain

$$\phi(x_n+h) - \phi(x_n-h) = 2h\phi'(x_n) + [\phi'''(\alpha_n) + \phi'''(\beta_n)]h^3/6.$$

From Problem 5 we have $y_{n+1} = y_{n-1} + 2hy_n'$, and thus e_{n+1}

$= \phi(x_n+h) - y_{n+1} = [\phi'''(\alpha_n) + \phi'''(\beta_n)]h^3/6$, since at each

step it is assumed that $\phi(x_n-h) = y_{n-1}$ and $\phi'(x_n) = y_n'$.

Hence the local discretization error is proportional to

h^3 plus possibly higher order terms.

7b.   The predictor formula is $y_{n+1} = y_{n-3} + (4h/3)[2y_n' - y_{n-1}'$

$+ 2 y_{n-2}']$ and the corrector formula is $y_{n+1} = y_{n-1} +$

$(h/3)[y_{n-1}' + 4 y_n' + y_{n+1}']$, where $y_n' = 1/2 - x_n + 2y_n$.

For the predictor formula we may use the values of y_n',

y_{n-1}' and y_{n-2}' calculated in Problem 2 and thus $y_{4p} = 1 +$

$(.4/3)[2(4.14422) - 3.48364 + 2(2.9428)] = 2.42539.$

Using this value for y_4 we find that $y_4' = 4.95077$ and

thus $y_{4c} = 1.59182 + (.1/3)[3.48364 + 4(4.14422) +$

$4.95078] = 2.425530$.

8b. Substituting $y = c \exp(Ax)$ in Eq.(5) we find that $y_{n+1} =$

$y_n + (h/24)(9Ay_{n+1} + 19 Ay_n - 5Ay_{n-1} + Ay_{n-2})$ or $(1 -$

$9Ah/24)y_{n+1} - (1 + 19Ah/24)y_n + 5Ahy_{n-1}/24 - Ahy_{n-2}/24 =$

0, where we may now set $\alpha = Ah/24$.

8c. Substituting $y_n = \lambda^n$ in Eq.(i) we have $(1-9\alpha)\lambda^{n+1} -$

$(1+19\alpha)\lambda^n + 5\alpha\lambda^{n-1} - \alpha\lambda^{n-2} = [(1-9\alpha)\lambda^3 - (1+19\alpha)\lambda^2 +$

$5\alpha\lambda - \alpha]\lambda^{n-2} = 0$.

8d. If we let $\alpha \to 0$ in Eq.(ii) we obtain $\lambda^3 - \lambda^2 = 0$ and

hence the roots are $\lambda_1 = 1$, $\lambda_2 = 0$, $\lambda_3 = 0$.

8e. Substituting $\lambda_1 = 1 + \lambda_{11}h$, $\lambda_1^2 = 1 + 2\lambda_{11}h + \lambda_{11}^2 h^2$, and

$\lambda_1^3 = 1 + 3\lambda_{11}h + 3\lambda_{11}^2 h^2 + \lambda_{11}^3 h^3$ in Eq.(ii) and neglecting

terms proportional to h^2 and h^3, we have $(1 - 9\alpha) \cdot$

$(1 + 3\lambda_{11}h) - (1 + 19\alpha)(1 + 2\lambda_{11}h) + 5\alpha(1 + \lambda_{11}h) - \alpha =$

$1 + 3\lambda_{11}h - 9\alpha - 1 - 2\lambda_{11}h - 19\alpha + 5\alpha - \alpha = 0$ (since $\alpha =$

$Ah/24$ we have neglected products of α and h).

Simplifying this last equation we obtain $\lambda_{11}h = 24\alpha = Ah$

and thus $\lambda_{11} = A$. Hence $\lambda_1 \simeq 1 + Ah$ which means $y_1 = \lambda_1^n$

$\simeq (1 + Ah)^n$. Now suppose we choose a point $x = x_n = nh$

and let $n \to \infty$ with $h \to 0$ so that $nh = x_n$ is fixed. Then

$y_1 \simeq (1+hA)^n = (1+Ax_n/n)^n \to \exp(Ax_n)$ as $h \to 0$.

Section 8.9, Page 440

1b.   The Euler formula is $x_{n+1} = x_n + hx_n'$, $y_{n+1} = y_n + hy_n'$

where $x_n' = 2x_n + t_n y_n$ and $y_n' = x_n y_n$, $n = 0,1,2...$ with x_0

$= 1, y_0 = 1$. Thus $x_1 = 1 + .1(2+0) = 1.2$, $y_1 = 1 +$

$.1[(1)(1)] = 1.1$ and $x_2 = 1.2 + .1[2.4 + (.1)(1.1)] =$

1.451, $y_2 = 1.1 + .1[(1.2)(1.1)] = 1.232$.

2.   If $x'(t) = f[t,x(t),y(t)]$, then the chain rule yields x''

$= \dfrac{\partial f}{\partial t} + \dfrac{\partial f}{\partial x} \dfrac{dx}{dt} + \dfrac{\partial f}{\partial y} \dfrac{dy}{dt}$, or $x'' = f_t(t,x,y) + f_x(t,x,y)x' +$

$f_y(t,x,y)y'$. A similar calculation yields y''. As shown

in Problem 3, the calculation of x'' and y'' is often less

complicated than the general formula would indicate.

3a.   Using the results of Problem 2, we have $x' = x+y+t$ and

thus $x'' = x'+y'+1$. Hence $x_1 = x_0 + h[x_0+y_0+t_0] +$

$(h^2/2)[x_0' + y_0' + 1]$. Substituting $x_0 = 1$ and $y_0 = 0$ we

obtain $x_1 = 1 + .2[1+0+0] + .02[1+4+1] = 1.32$. Note that

we also needed $y_0' = 4x_0 - 2y_0$. In a similar fashion $y_1 =$

$y_0 + h[4x_0 - 2y_0] + (h^2/2)[4x_0' - 2y_0'] = 1 + .2[4-0] +$

$.02[4-8] = .72$.

6.   If we let $y = x'$, then $y' = x''$ and thus we obtain the

sytem $x' = y$ and $y' = t-3x-t^2 y$, with $x(0) = 1$ and $y(0) =$

$x'(0) = 2$. The Euler formula is $x_{n+1} = x_n + hx_n'$ and y_{n+1}

$= y_n + hy_n'$ and hence $x_1 = 1 + .1(2) = 1.2$, $y_1 = 2 +$

$.1(-3) = 1.7$ and $x_2 = 1.2 + .1(1.7) = 1.37$, $y_2 = 1.7 +$

$.1[.1-3.6-(.01)(1.7)] = 1.3483$.

CHAPTER 9

Section 9.1, Page 454

1a. Solutions of the D.E. for x

and y are $x = Ae^{-t}$ and y =

Be^{-2t} respectively. The I.C.

$x(0) = 4$ and $y(0) = 2$ yield

A = 4 and B = 2, so $x = 4e^{-t}$,

and $y = 2e^{-2t}$.

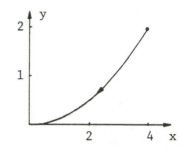

Solving the first equation for e^{-t} and then substituting

into the second yields $y = 2[x/4]^2 = x^2/8$, which is a

parabola. From the original D.E., or from the parametric

solutions, we find that $0 < x \le 4$ and $0 < y \le 2$ for $t \ge 0$

and thus only the portion of the parabola shown is the

trajectory, with the direction of motion indicated.

1c. Utilizing the approach indicated in Eq.(11), we have

$dy/dx = -x/y$, which separates into $xdx + ydy = 0$.

Integration then yields $x^2 + y^2 = c^2$, where $c^2 = 16$ for

both sets of I.C. The direction of motion can be found

from the original D.E. and is counterclockwise for both

I.C. To obtain the parametric equations, we write the

system in the form $\frac{d}{dt} \begin{pmatrix} x \\ y \end{pmatrix} = \begin{pmatrix} 0 & -1 \\ 1 & 0 \end{pmatrix} \begin{pmatrix} x \\ y \end{pmatrix}$, which has the

characteristic equation $\begin{vmatrix} -r & -1 \\ 1 & -r \end{vmatrix} = r^2 + 1 = 0$, or

$r = \pm i$. Following the procedures of Section 7.6, we find that one solution of the above system is $\begin{pmatrix} 1 \\ -i \end{pmatrix} e^{it} =$

$\begin{pmatrix} \cos t + i \sin t \\ \sin t - i \cos t \end{pmatrix}$ and thus two real solutions are

$\underset{\sim}{u}(t) = \begin{pmatrix} \cos t \\ \sin t \end{pmatrix}$ and $\underset{\sim}{v}(t) = \begin{pmatrix} \sin t \\ -\cos t \end{pmatrix}$. The general

solution of the system is then $\begin{pmatrix} x \\ y \end{pmatrix} = c_1 \underset{\sim}{u}(t) + c_2 \underset{\sim}{v}(t)$ and

hence the first I.C. yields $c_1 = 4$, $c_2 = 0$, or $x =$

4 cos t, y = 4 sin t. The second I.C. yields $c_1 = 0$,

$c_2 = -4$, or x = -4 sin t, y = 4 cos t. Note that both

these parametric representations satisfy the form of the

trajectories found in the first part of this problem.

2c. The critical points are given by the solutions of

x(1-x-y) = 0 and y(1/2 - y/4 - 3x/4) = 0. The solutions

corresponding to either x = 0 or y = 0 are seen to be x =

0, y = 0; x = 0, y = 2; x = 1, y = 0. In addition, there

is a solution corresponding to the intersection of the

lines 1 - x - y = 0 and 1/2 - y/4 - 3x/4 = 0 which is the

point x = 1/2, y = 1/2. Thus

the critical points are

(0,0), (0,2), (1,0), and

(1/2,1/2).

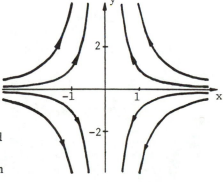

3b. Integrating each of the D.E.

yields x = Ae^{-t}, y = Be^{2t} and

thus y = A^2B/x^2, A ≠ 0. From

the D.E. we have $dx/dt < 0$ and $dy/dt > 0$ in the first quadrant and thus the motion is as shown. Each quadrant must be done separately for this problem and hence the results are as shown.

4.   We know that $\phi'(t) = F[\phi(t), \psi(t)]$ and $\psi'(t) = G[\phi(t), \psi(t)]$ for $\alpha < t < \beta$. By direct substitution we have

$\Phi'(t) = \phi'(t-s) = F[\phi(t-s), \psi(t-s)] = F[\Phi(t), \Psi(t)]$ and

$\Psi'(t) = \psi'(t-s) = G[\phi(t-s), \psi(t-s)] = G[\Phi(t), \Psi(t)]$ for

$\alpha < t-s < \beta$ or $\alpha+s < t < \beta +s$.

6.   If $F(x_0, y_0, t) = G(x_0, y_0, t) = 0$ for all t then the conclusion is obvious. In a region where $F(x,y,t) \neq 0$ we may write

$$\frac{dy}{dx} = \frac{dy/dt}{dx/dt} = \frac{G(x,y,t)}{F(x,y,t)} = \frac{G(x,y,t)/[F^2(x,y,t) + G^2(x,y,t)]^{1/2}}{F(x,y,t)/[F^2(x,y,t) + G^2(x,y,t)]^{1/2}}$$

$$= \frac{A(x,y)}{B(x,y)} .$$

This is an autonomous D.E. which yields a one-parameter family of solutions independent of s. This family is the set of trajectories of the original system. A similar argument holds near points where $F(x,y,t) = 0$ but $G(x,y,t) \neq 0$ if we consider $dx/dy = F(x,y,t)/G(x,y,t)$.

7.   Letting $x = \theta$ and $y = dx/dt$, then the D.E. can be written as the system $dx/dt = y$, $dy/dt = -(g/\ell) \sin x$. Using the approach of Eq.(11), we find that this system has $dy/dx = (-g/\ell)\sin x/y$, which separates into $y \, dy = (-g/\ell)\sin x \, dx$.

Integrating both sides yields $y^2/2 = (g/\ell) \cos x +$

c_1. Setting $c_1 = c - g/\ell$ then yields the desired form.

10. Suppose that $t_1 > t_0$. Let $s = t_1 - t_0$. Since the system

is autonomous, the result of Problem 4 with s replaced by

$-s$ shows that $x = \phi_1(t+s)$ and $y = \psi_1(t+s)$ generates the

same trajectory (C_1) as $x = \phi_1(t)$ and $y = \psi_1(t)$. But at

$t = t_0$ we have $x = \phi_1(t_0+s) = \phi_1(t_1) = x_0$ and $y =$

$\psi_1(t_0+s) = \psi_1(t_1) = y_0$. Thus the solution $x = \phi_1(t+s)$,

$y = \psi_1(t+s)$ satisfies <u>exactly</u> the same initial conditions

as the solution $x = \phi_0(t)$, $y = \psi_0(t)$ which generates the

trajectory C_0. Hence C_0 and C_1 are the same.

11. From the existence and uniqueness theorem we know that if

the two solutions $x = \phi(t)$, $y = \psi(t)$ and $x = x_0$, $y = y_0$

satisfy $\phi(a) = x_0$, $\psi(a) = y_0$ and $x = x_0$, $y = y_0$ at $t = a$,

then these solutions are identical. Hence $\phi(t) = x_0$ and

$\psi(t) = y_0$ for all t contradicting the fact that the

trajectory generated by $[\phi(t), \psi(t)]$ started at a

noncritical point.

12. By direct substitution $\Phi'(t) = \phi'(t+T) =$

$F[\phi(t+T), \psi(t+T)] = F[\Phi(t), \Psi(t)]$ and $\Psi'(t) = \psi'(t+T) =$

$G[\phi(t+T), \psi(t+T)], = G[\Phi(t), \Psi(t)]$. Furthermore $\Phi(t_0) =$

x_0 and $\Psi(t_0) = y_0$. Thus by the existence and uniqueness

theorem $\Phi(t) = \phi(t)$ and $\Psi(t) = \psi(t)$ for all t.

Section 9.2, Page 467

For Problems 1 through 16, once the eigenvalues have been found, Table 9.1 will for the most part quickly yield the type of critical point and the stability. In all cases it can be easily verified that $\underset{\sim}{A}$ is nonsingular.

1. The eigenvalues are found from the equation $\det(\underset{\sim}{A}-r\underset{\sim}{I}) = 0$. Substituting the values for $\underset{\sim}{A}$ we have $\begin{vmatrix} 3-r & -2 \\ 2 & -2-r \end{vmatrix} = r^2 - r + 2 = 0$ and thus the eigenvalues are $r_1 = -1$ and $r_2 = 2$. Since the eigenvalues differ in sign, the critical point is a saddle point and is unstable.

4. Again the eigenvalues are given by $\begin{vmatrix} 1-r & -4 \\ 4 & -7-r \end{vmatrix} = r^2 + 6r + 9 = 0$ and thus $r_1 = r_2 = -3$. To determine whether the critical point is a proper or improper node we must determine whether there are two or one independent eigenvectors. In this case the eigenvectors are solutions of $\begin{pmatrix} 4 & -4 \\ 4 & -4 \end{pmatrix}\begin{pmatrix} \xi_1 \\ \xi_2 \end{pmatrix} = \begin{pmatrix} 0 \\ 0 \end{pmatrix}$ and hence there is just one eigenvector. Thus $(0,0)$ is an improper node which is asymptotically stable since the eigenvalues are negative. If we had found that there were two independent eigenvectors then $(0,0)$ would have been a proper node, as indicated in Case 3a.

7. In this case $\det(\underset{\sim}{A} - r\underset{\sim}{I}) = r^2 - 2r + 5$ and thus the eigenvalues are $r_{1,2} = 1 \pm 2i$. Since these are complex, with positive real part we conclude that the critical point is a spiral point and unstable.

10. Again, det $(\underset{\sim}{A} - r\underset{\sim}{I}) = r^2 + 9$ and thus we have $r_{1,2} = \pm 3i$.

Since the eigenvalues are pure imaginary the critical

point is a center, which is stable.

13. If we let $\underset{\sim}{x} = \underset{\sim}{x}^0 + \underset{\sim}{u}$ then $\underset{\sim}{x}' = \underset{\sim}{u}'$ and thus the system

becomes $\underset{\sim}{u}' = \begin{pmatrix} 1 & 1 \\ 1 & -1 \end{pmatrix} \underset{\sim}{x}^0 + \begin{pmatrix} 1 & 1 \\ 1 & -1 \end{pmatrix} \underset{\sim}{u} - \begin{pmatrix} 2 \\ 0 \end{pmatrix}$

which will be in the form of Eq.(2) if $\begin{pmatrix} 1 & 1 \\ 1 & -1 \end{pmatrix} \underset{\sim}{x}^0 = \begin{pmatrix} 2 \\ 0 \end{pmatrix}$.

Using row operations, this last set of equations is

equivalent to $\begin{pmatrix} 1 & 1 \\ 0 & -2 \end{pmatrix} \underset{\sim}{x}^0 = \begin{pmatrix} 2 \\ -2 \end{pmatrix}$ and thus $x_0 = 1$ and $y_0 = $

1. Since $\underset{\sim}{u}' = \begin{pmatrix} 1 & 1 \\ 1 & -1 \end{pmatrix} \underset{\sim}{u}$ has $(0,0)$ as the critical point,

we conclude that $(1,1)$ is the critical point of the

original system. As in the earlier problems, the

eigenvalues are given by $\begin{vmatrix} 1-r & 1 \\ 1 & -1-r \end{vmatrix} = r^2 - 2 = 0$ and thus

$r_{1,2} = \pm \sqrt{2}$. Hence the critical point $(1,1)$ is an

unstable saddle point.

17. The equivalent system is $dx/dt = y$, $dy/dt = - (k/m)x - $

$(c/m)y$, which is written in the form of Eq.(2) as

$\dfrac{d}{dt} \begin{pmatrix} x \\ y \end{pmatrix} = \begin{pmatrix} 0 & 1 \\ -k/m & -c/m \end{pmatrix} \begin{pmatrix} x \\ y \end{pmatrix}$. The point $(0,0)$ is clearly a

critical point, and since $\underset{\sim}{A}$ is nonsingular, it is the

only one. The characteristic equation is $r^2 + (c/m)r + $

$k/m = 0$ so $r_1, r_2 = [- c \pm (c^2 - 4km)^{1/2}]/2m$. In the

underdamped case $c^2 - 4km < 0$, and the characteristic

roots are complex with negative real parts. Thus the

critical point $(0,0)$ is an asymptotically stable spiral

point. In the overdamped case $c^2 - 4km > 0$, and the

characteristic roots are real, unequal, and negative.

Thus the critical point $(0,0)$ is an asymptotically stable

improper node. In the critically damped case $c^2 - 4km =$

0, and the characteristic roots are equal and negative.

As indicated in the solution to Problem 4, to determine

whether this is an improper or proper node we must

determine whether there are one or two linearly

independent eigenvectors. The eigenvectors satisfy the

equations $\begin{pmatrix} c/2m & 1 \\ -k/m & -c/2m \end{pmatrix}\begin{pmatrix} \xi_1 \\ \xi_2 \end{pmatrix} = \begin{pmatrix} 0 \\ 0 \end{pmatrix}$, which have just one

solution if $c^2 - 4km = 0$. Thus the critical point $(0,0)$

is a stable improper node.

18a. If $\underset{\sim}{A}$ has one zero eigenvalue then for $r = 0$ we have

   $\det (\underset{\sim}{A} - r\underset{\sim}{I}) = \det \underset{\sim}{A} = 0$. Hence $\underset{\sim}{A}$ is singular which means

   $\underset{\sim}{A}\ \underset{\sim}{x} = \underset{\sim}{0}$ has infinitely many solutions and consequently

   there are infinitely many critical points.

18b. From Chapter 7, the solution is $\underset{\sim}{x}(t) = c_1 \underset{\sim}{\xi}^{(1)} + c_2 \underset{\sim}{\xi}^{(2)} .$
   $e^{r_2 t}$, which can be written in scalar form as $x_1 =$

   $c_1 \xi_1^{(1)} + c_2 \xi_1^{(2)} e^{r_2 t}$ and $x_2 = c_1 \xi_2^{(1)} + c_2 \xi_2^{(2)} e^{r_2 t}$.

   Assuming $\xi_1^{(2)} \neq 0$, the first equation can be solved for

   $c_2 e^{r_2 t}$, which is then substituted into the second

   equation to yield $x_2 = c_1 \xi_2^{(1)} + [\xi_2^{(2)}/\xi_1^{(2)}][x_1 - c_1 \xi_1^{(1)}]$.

   These are straight lines parallel to the vector

   $\underset{\sim}{\xi}^{(2)}$. Note that the family of lines is independent of c_2.

If $\xi_1^{(2)} = 0$, then the lines are vertical.

21b. Eq.(i) can be written in scalar form as $dx/dt = ax + by$

and $dy/dt = cx + dy$, which then yields Eq.(iii). Ignoring

the middle quotient in Eq.(iii), we can rewrite that

equation as $(cx + dy)dx - (ax + by)dy = 0$, which is exact

since $d = -a$.

21c. Integrating $\phi_x = cx + dy$ we obtain $\phi = cx^2/2 + dxy +$

$g(y)$, and thus $dx + dg/dy = -ax - by$, or $dg/dy = -by$.

Hence $cx^2/2 + dxy - by^2/2 = k/2$ is the solution to

Eq.(iii). The quadratic equation $Ax^2 + Bxy + Cy^2 = 0$ is

an ellipse provided $B^2 - 4AC < 0$. Hence for our problem

if $4d^2 + 4bc < 0$ then Eq.(iv) is an ellipse. Using

$a + d = 0$ we have $d^2 = -ad$ and hence $-ad + bc < 0$ or

$ad - bc > 0$, which is true by Eqs.(ii). Thus Eq.(iv) is an

ellipse under the conditions of Eqs.(ii).

22. The given system can be written as $\dfrac{d}{dt}\begin{pmatrix} x \\ y \end{pmatrix} = \begin{pmatrix} a & b \\ c & d \end{pmatrix}\begin{pmatrix} x \\ y \end{pmatrix}$

and thus the eigenvalues are given by $r^2 - (a+d)r + ad -$

$bc = 0$. Using the given definitions we rewrite this as

$r^2 - pr + q = 0$ and thus $r_{1,2} = (p \pm \sqrt{p^2-4q})/2 =$

$(p \pm \sqrt{\Delta})/2$. The results are now obtained using Table

9.1.

Section 9.3, Page 482

In Problems 1 through 10, write the system in the form of
Eq.(7). Then if $g(0) = 0$ we may conclude that $(0,0)$ is a

critical point. In addition, if g satisfies Eq.(9) or
Eq.(10), then the system is almost linear. In this case the
linear system, Eq.(8), will determine, in most cases, the type
and stability of the critical point (0,0) of the almost linear
system. These results are summarized in Table 9.2.

1.   In this case the system can be written as

$$\frac{d}{dt}\begin{pmatrix} x \\ y \end{pmatrix} = \begin{pmatrix} 1 & -1 \\ 3 & -2 \end{pmatrix}\begin{pmatrix} x \\ y \end{pmatrix} + \begin{pmatrix} xy \\ -xy \end{pmatrix}$$

and thus $\underset{\sim}{A} = \begin{pmatrix} 1 & -1 \\ 3 & -2 \end{pmatrix}$ and $\underset{\sim}{g} = \begin{pmatrix} xy \\ -xy \end{pmatrix}$. Since $\underset{\sim}{g}(0) = \begin{pmatrix} 0 \\ 0 \end{pmatrix}$

we conclude that (0,0) is a critical point. Following

the procedure of Example 1, we let $x = r\cos\theta$ and $y =$

$r\sin\theta$ and thus $g_1(x,y)/r = -g_2(x,y)/r = \dfrac{r^2\cos\theta\,\sin\theta}{r} \to 0$

as $r \to 0$ and thus the system is almost linear. Since

$\det(\underset{\sim}{A}-r\underset{\sim}{I}) = r^2 + r + 1$, we find that the eigenvalues are

$r_{1,2} = -1/2 \pm \sqrt{3}i/2$ and thus (0,0) is an asymptotically

stable spiral point.

2.   We have $\dfrac{d}{dt}\begin{pmatrix} x \\ y \end{pmatrix} = \begin{pmatrix} 1 & 0 \\ 0 & 1 \end{pmatrix}\begin{pmatrix} x \\ y \end{pmatrix} + \begin{pmatrix} x^2+ y^2 \\ -xy \end{pmatrix}$ and since

$\underset{\sim}{g}(0) = \begin{pmatrix} 0 \\ 0 \end{pmatrix}$ we conclude that (0,0) is a critical point.

Letting $x = r\cos\theta$ and $y = r\sin\theta$ we see that $g_1(x,y)/r$

$= (r^2\cos^2\theta + r^2\sin^2\theta)/r \to 0$ and that $g_2(x,y)/r =$

$r^2\cos\theta\,\sin\theta /r \to 0$ as $r \to 0$. Hence the system is almost

linear. The eigenvalues for the corresponding linear

system are found to be $r_1 = 1$ and $r_2 = 1$. For the linear

system we could determine whether (0,0) was a proper or

improper node. However, for the almost linear system the

repeated eigenvalues will be perturbed and thus we cannot

determine whether (0,0) is a proper node, improper node

or a spiral point (if the perturbed values are complex).

In any case, since r_1 and r_2 are positive, the system

will be unstable.

8.   The given system can be written as

$$\frac{d}{dt}\begin{pmatrix} x \\ y \end{pmatrix} = \begin{pmatrix} 0 & 1 \\ 0 & 1 \end{pmatrix} \begin{pmatrix} x \\ y \end{pmatrix} + \begin{pmatrix} 1 - e^{-x} \\ -\sin x \end{pmatrix}.$$

In this case, though, $\underset{\sim}{A} = \begin{pmatrix} 0 & 1 \\ 0 & 1 \end{pmatrix}$ is singular and $\underset{\sim}{g} =$

$\begin{pmatrix} 1 - e^{-x} \\ -\sin x \end{pmatrix}$ does not satisfy Eq.(10) and thus we may not

proceed. However, if we consider the Taylor series for

e^{-x} and for sin x, we see that $1 - x - e^{-x} = -\frac{x^2}{2!} + \frac{x^3}{3!} +$

... and $-\sin x + x = \frac{x^3}{3!} - \frac{x^5}{5!} + \ldots$ and hence if we let x =

r cos θ we may conclude that $(1 - x - e^{-x})/r \to 0$ and

$(-\sin x + x)/r \to 0$ as $r \to 0$. Using this information we now

add and subtract x to the first equation to obtain $\frac{dx}{dt} =$

$x + y + 1 - x - e^{-x}$ and similarly for the second equation

to obtain $\frac{dy}{dt} = -x + y - \sin x + x$. This equivalent

system can now be written as

$$\frac{d}{dt}\begin{pmatrix} x \\ y \end{pmatrix} = \begin{pmatrix} 1 & 1 \\ -1 & 1 \end{pmatrix} \begin{pmatrix} x \\ y \end{pmatrix} + \begin{pmatrix} 1 - x - e^{-x} \\ -\sin x + x \end{pmatrix}. \quad \text{Clearly,}$$

(0,0) is a critical point and the system is almost linear

by the above analysis. The linear system, with $\underset{\sim}{A} = \begin{pmatrix} 1 & 1 \\ -1 & 1 \end{pmatrix}$,

has the eigenvalues $r_{1,2} = 1 \pm i$ and hence the almost

linear system has an unstable spiral point at (0,0).

11c. The critical points are the solutions of $x(1-x-y) = 0$ and

$y(3-x-2y) = 0$. Solutions are $x = 0$, $y = 0$; $x = 0$,

$3 - 2y = 0$ which gives $y = 3/2$; $y = 0$ and $1 - x = 0$ which

gives $x = 1$; and $1 - x - y = 0$, $3 - x - 2y = 0$ which

gives $x = -1$, $y = 2$. Thus the critical points are $(0,0)$,

$(0, 3/2)$, $(1,0)$, and $(-1,2)$. For the critical point

$(0,0)$ the D.E. is already in the form of an almost linear

system; and the corresponding linear system is $dx/dt = x$,

$dy/dt = 3y$ which has the eigenvalues $r_1 = 1$ and $r_2 = 3$.

Thus the critical point $(0,0)$ is an unstable improper

node. Each of the other three critical points is

dealt with in the same manner; we consider only the

critical point $(-1, 2)$. In order to translate this

critical point to the origin we set $x(t) = - 1 + u(t)$,

$y(t) = 2 + v(t)$ and substitute in the D.E. to obtain

$$du/dt = - 1 + u - (-1+u)^2 - (-1+u)(2+v) = u + v - u^2 - uv$$

and

$$dv/dt = 3(2+v) - (-1+u)(2+v) - 2(2+v)^2 = -2u - 4v - uv - 2v^2.$$

Writing this in the form of Eq.(7) we find that $\underset{\sim}{A} =$

$\begin{pmatrix} 1 & 1 \\ -2 & -4 \end{pmatrix}$ and $\underset{\sim}{g} = - \begin{pmatrix} u^2 + uv \\ uv + v^2 \end{pmatrix}$ which is an almost linear

system. The eigenvalues of the corresponding linear

system are $r = (-3 \pm \sqrt{9 + 8})/2$ and hence the critical

point $(-1,2)$, of the original system, is an unstable

saddle point.

13a. The system is $\dfrac{d}{dt}\begin{pmatrix} x \\ y \end{pmatrix} = \begin{pmatrix} 1 & 0 \\ 0 & -2 \end{pmatrix}\begin{pmatrix} x \\ y \end{pmatrix} + \begin{pmatrix} 0 \\ x^2 \end{pmatrix}$ and

thus is almost linear using the procedures outlined in

the earlier problems. The corresponding linear system

has the eigenvalues $r_1 = 1$, $r_2 = -2$ and thus $(0,0)$ is an

unstable saddle point.

13b. The trajectories of the linear system are the solutions

of $dx/dt = x$ and $dy/dt = -2y$ and thus $x(t) = c_1 e^t$ and

$y(t) = c_2 e^{-2t}$. The sketch these, solve the first

equation for e^t and substitute into the second to obtain

$y = c_1^2 c_2/x^2$, $c_1 \neq 0$. Several trajectories are shown in

the figure. Since $x(t) = c_1 e^t$, we must pick $c_1 = 0$ for

$x \to 0$ as $t \to \infty$. Thus $x = 0$,

$y = c_2 e^{-2t}$ (the vertical axis)

is the only trajectory for

which $x \to 0$, $y \to 0$ as $t \to \infty$.

13c. For $x \neq 0$ we have $dy/dx = (dy/dt)/(dx/dt) = (-2y+x^3)/x$.

This is a linear equation, and the general solution is

$y = x^3/5 + k/x^2$, where k is an arbitrary constant. In

addition the system of equations has the solution $x = 0$,

$y = Be^{-2t}$. Any solution with its initial point on the

y-axis (x=0) is given by the latter solution. The

trajectories corresponding to these solutions approach

the origin as $t \to \infty$. The trajectory that passes through

the origin and divides the

family of curves is given by

k = 0, namely $y = x^3/5$. This

trajectory corresponds to

the trajectory y = 0 for the

linear problem. Several

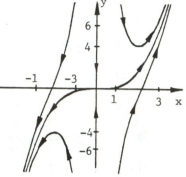

trajectories are sketched in the figure.

16. The equation of the trajectories was found in Problem 7

of Section 9.1.

17. Setting c = 0 in Eq.(15) and multiplying by $m\ell^2$ we obtain

$m\ell^2 d^2\theta/dt^2 + mg\ell \sin\theta = 0$. Considering $d\theta/dt$ as a

function of θ and using the chain rule we have

$$\frac{d}{dt}\left(\frac{d\theta}{dt}\right) = \frac{d}{d\theta}\left(\frac{d\theta}{dt}\right)\frac{d\theta}{dt} = \frac{1}{2}\frac{d}{d\theta}\left(\frac{d\theta}{dt}\right)^2 .$$

Thus $(1/2)m\ell^2 d[(d\theta/dt)^2]/d\theta = -mg\ell \sin\theta$. Now integrate

both sides from α to θ where $d\theta/dt = 0$ at $\theta = \alpha$:

$(1/2)m\ell^2(d\theta/dt)^2 = mg\ell(\cos\theta - \cos\alpha)$. Thus $(d\theta/dt)^2 =$

$(2g/\ell)(\cos\theta - \cos\alpha)$. Since we are releasing the

pendulum with zero velocity from a positive angle α, the

angle θ will initially be decreasing so $d\theta/dt < 0$. If we

restrict attention to the range of θ from $\theta = \alpha$ to $\theta = 0$, we can assert $d\theta/dt = -\sqrt{2g/\ell}\,\sqrt{\cos\theta - \cos\alpha}$. Solving

for dt gives $dt = -\sqrt{\ell/2g}\ d\theta/\sqrt{\cos\theta - \cos\alpha}$.

Since there is no damping, the pendulum will swing from

its initial angle α through 0 to $-\alpha$, then back through

0 again to the angle α in one period. It follows that

$\theta(T/4) = 0$. Integrating the last equation and noting

that as t goes from 0 to T/4, θ goes from α to 0 yields

$T/4 = -\sqrt{\ell/2g}\ \int_{\alpha}^{0} (1/\sqrt{\cos\theta - \cos\alpha})d\theta$. We obtain the

elliptic integral by making the change of variables given

in the text.

18b. Using the hint, the system can be written as

$$\frac{d}{dt}\begin{pmatrix} x \\ y \end{pmatrix} = \begin{pmatrix} 0 & 1 \\ -g'(0) & -c(0) \end{pmatrix}\begin{pmatrix} x \\ y \end{pmatrix} - \begin{pmatrix} 0 \\ -g''(\xi_1)x^2/2 - c'(\xi_2)xy \end{pmatrix}.$$

where $0 < \xi_1, \xi_2 < x$, from which the results follow.

19a. If both species are to coexist, then neither X = 0 or Y =

0. Such a critical point will correspond to the

intersection of the two lines $\varepsilon_1 - \sigma_1 x - \alpha_1 y = 0$, $\varepsilon_2 - $

$\sigma_2 y - \alpha_2 x = 0$ if they intersect. The necessary and

sufficient condition that there be a unique solution of

this linear nonhomogeneous system is that the determinant

of the coefficients of x and y be nonzero, namely, $\sigma_1\sigma_2 \neq$

$\alpha_1\alpha_2$. The fact that α_1, α_2, σ_1, σ_2, ε_1, ε_2 are all

positive constants and $\varepsilon_1/\sigma_1 < \varepsilon_2/\alpha_2$, $\varepsilon_2/\sigma_2 < \varepsilon_1/\alpha_1$

implies that $\alpha_2/\sigma_1 < \varepsilon_2/\varepsilon_1 < \sigma_2/\alpha_1$ or $\alpha_1\alpha_2 < \sigma_1\sigma_2$.

Solving the system yields the critical point $X = (\varepsilon_1\sigma_2 - \varepsilon_2\alpha_1)/(\sigma_1\sigma_2 - \alpha_1\alpha_2)$ and $Y = (\varepsilon_2\sigma_1 - \varepsilon_1\alpha_2)/(\sigma_1\sigma_2 - \alpha_1\alpha_2)$, which are both positive using the above inequalities.

19b. Substituting $x = X + u$, $y = Y + v$, in the system of D.E. yields

$$du/dt = (X + u)[\varepsilon_1 - \sigma_1(X + u) - \alpha_1(Y + v)]$$
$$= (X + u)(- \sigma_1 u - \alpha_1 v)$$
$$= - \sigma_1 Xu - \alpha_1 Xv - \sigma_1 u^2 - \alpha_1 uv ,$$
$$dv/dt = (Y + v)[\varepsilon_2 - \sigma_2(Y + v) - \alpha_2(X + u)]$$
$$= (Y + v)(- \sigma_2 v - \alpha_2 u)$$
$$= - \alpha_2 Yu - \sigma_2 Yv - \alpha_2 uv - \sigma_2 v^2 .$$

Note that we have used the fact that $\varepsilon_1 - \sigma_1 X - \alpha_1 Y = 0$ and $\varepsilon_2 - \sigma_2 Y - \alpha_2 X = 0$. This system of equations has a critical point at $u = 0$, $v = 0$. Letting $u = r \cos\theta$, $v = r \sin\theta$ it is easily shown that u^2/r, uv/r, $v^2/r \to 0$ as $r \to 0$ so the transformed system is almost linear in the neighborhood of the critical point $(0,0)$.

19c. The corresponding linear system is

$$\frac{d}{dt} \begin{pmatrix} u \\ v \end{pmatrix} = \begin{pmatrix} -\sigma_1 X & -\alpha_1 X \\ -\alpha_2 Y & -\sigma_2 Y \end{pmatrix} \begin{pmatrix} u \\ v \end{pmatrix} \quad \text{which has the}$$

characteristic equation $r^2 + (\sigma_1 X + \sigma_2 Y)r + (\sigma_1\sigma_2 - \alpha_1\alpha_2)XY = (r-r_1)(r-r_2) = 0$. Note that $r_1 + r_2 = - (\sigma_1 X + \sigma_2 Y)$ and $r_1 r_2 = (\sigma_1\sigma_2 - \alpha_1\alpha_2)XY$. If r_1 and r_2 are real it follows, since $(\sigma_1\sigma_2 - \alpha_1\alpha_2)XY > 0$, from the second equation that they have the same sign and from the

first equation that each is negative. To show that r_1

and r_2 are real and not equal we must show that the

discriminant of the quadratic is positive; namely,

$(\sigma_1 X + \sigma_2 Y)^2 - 4(\sigma_1 \sigma_2 - \alpha_1 \alpha_2)XY > 0$. This can be

written as $(\sigma_1 X - \sigma_2 Y)^2 + 4 \alpha_1 \alpha_2 XY$, and since $X > 0$, $Y > 0$

we see that the discriminant is positive. Thus we

conclude that (X,Y) is an asymptotically stable improper

node.

Section 9.4, Page 498

1a. The critical points are found by setting $dx/dt = 0$ and

$dy/dt = 0$ and thus we need to solve $x(1-x + y/2) = 0$ and

$y(5/2 - 3y/2 + x/4) = 0$. The first yields $x = 0$ or $y =$

$2x - 2$ and the second yields $y = 0$ or $y = x/6 + 5/3$.

Thus we find the critical points $(0,0)$, $(1,0)$, $(0,5/3)$

and $(2,2)$. The last point is the intersection of the two

straight lines, which will be used again in part c.

1b. For $(0,0)$ the linearized system is $x' = x$ and $y' = 5y/2$,

which has the eigenvalues $r_1 = 1$ and $r_2 = 5/2$. Thus the

origin is an unstable improper node. For $(2,2)$ we let

$x = u + 2$ and $y = v + 2$ in the given system to find

(since $x' = u'$ and $y' = v'$) that

$du/dt = (u+2)[1- (u+2) + (v+2)/2] = (u+2)(-u+v/2)$ and

$dv/dt = (v+2)[5/2 - 3(v+2)/2 + (u+2)/4] = (v+2)(u/4 - 3v/2)$.

Hence the linearized equations are $\begin{pmatrix} u \\ v \end{pmatrix}' = \begin{pmatrix} -2 & 1 \\ 1/2 & -3 \end{pmatrix}\begin{pmatrix} u \\ v \end{pmatrix}$

which has the eigenvalues $r_{1,2} = (-5 \pm \sqrt{3})/2$. Since

these are both negative we conclude that (2,2) is an

asymptotically stable improper node. In a similar

fashion for (1,0) we let x = u + 1 and y = v to obtain

the linearized system $\begin{pmatrix} u \\ v \end{pmatrix}' = \begin{pmatrix} -1 & 1/2 \\ 0 & 11/4 \end{pmatrix}\begin{pmatrix} u \\ v \end{pmatrix}$. This has

$r_1 = -1$ and $r_2 = 11/4$ as eigenvalues and thus (1,0) is

an unstable saddle point. Likewise, for (0,5/3) we let

x = u, y = v + 5/3 to find $\begin{pmatrix} u \\ v \end{pmatrix}' = \begin{pmatrix} 11/6 & 0 \\ 5/12 & -5/2 \end{pmatrix}\begin{pmatrix} u \\ v \end{pmatrix}$ as the

corresponding linear system. Thus $r_1 = 11/6$ and $r_2 = -5/2$

and thus (0,5/3) is an unstable saddle point.

1c.   To sketch the required trajectories, we must find the

eigenvectors for each of the linearized systems and then

analyze the behavior of the linear solution near the

critical point. Using this approach we find that the

solution near (0,0) has the form $\begin{pmatrix} x \\ y \end{pmatrix} = c_1 \begin{pmatrix} 1 \\ 0 \end{pmatrix} e^t +$

$c_2 \begin{pmatrix} 0 \\ 1 \end{pmatrix} e^{5t/2}$ and thus the origin is approached only for

large negative values of t. In this case e^t dominates

$e^{5t/2}$ and hence in the neighborhood of the origin all

trajectories are tangent to the x-axis except for one

pair ($c_1 = 0$) that lies along the y-axis.

For (2,2) we find the eigenvector corresponding to r =

$(-5 + \sqrt{3})/2 = -1.63$ is given by $(1 - \sqrt{3})\xi_1/2 + \xi_2 = 0$ and thus $\left(\begin{smallmatrix} 1 \\ (\sqrt{3}-1)/2 \end{smallmatrix}\right) = \left(\begin{smallmatrix} 1 \\ .37 \end{smallmatrix}\right)$ is one eigenvector. For r =

$(-5 -\sqrt{3})/2 = -3.37$ we have $(1 +\sqrt{3})\xi_1/2 + \xi_2 = 0$ and thus

$\left(\begin{smallmatrix} 1 \\ -(\sqrt{3}+1)/2 \end{smallmatrix}\right) = \left(\begin{smallmatrix} 1 \\ -1.37 \end{smallmatrix}\right)$ is the second eigenvector. Hence

the linearized solution is

$$\begin{pmatrix} u \\ v \end{pmatrix} = c_1 \begin{pmatrix} 1 \\ .37 \end{pmatrix} e^{-1.63t} + c_2 \begin{pmatrix} 1 \\ -1.37 \end{pmatrix} e^{-3.37t}.$$

For large positive values of t the first term is the

dominant one and thus we conclude that all trajectories

but two approach (2,2) tangent to the straight line with

slope .37. If $c_1 = 0$, we see that there are exactly two

($c_2 > 0$ and $c_2 < 0$) trajectories that lie on the straight

line with slope -1.37.

In a similar fashion, we find the linearized solutions

near (1,0) and (0,5/3) to be, respectively,

$$\begin{pmatrix} u \\ v \end{pmatrix} = c_1 \begin{pmatrix} 1 \\ 0 \end{pmatrix} e^{-t} + c_2 \begin{pmatrix} 1 \\ 15/2 \end{pmatrix} e^{11t/4} \quad \text{and}$$

$$\begin{pmatrix} u \\ v \end{pmatrix} = c_1 \begin{pmatrix} 0 \\ 1 \end{pmatrix} e^{-5t/2} + c_2 \begin{pmatrix} 1 \\ 5/52 \end{pmatrix} e^{11t/6} \quad ,$$

which, along with the

above analysis, yields

the sketch shown:

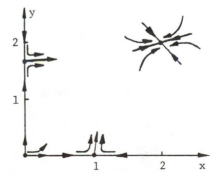

From the above sketch, it appears that $(x,y) \to (2,2)$ as

$t \to \infty$ as long as (x,y) starts in the first quadrant.

To ascertain this, we need to prove that x and y cannot become unbounded as t → ∞. From the given system, we can observe that, since $x > 0$ and $y > 0$, that dx/dt and dy/dt have the same sign as the quantities $1 - x + y/2$ and $5/2 - 3y/2 + x/4$ respectively. If we set these quantities equal to zero we get the straight lines $y = 2x - 2$ and $y = x/6 + 5/3$, which divide the first quadrant into the four sectors shown below. The signs of x' and y' are indicated, from which it can be concluded that x and y must remain bounded [and in fact approach $(2,2)$] as t → ∞. The discussion leading up to Fig.9.25b is also useful here.

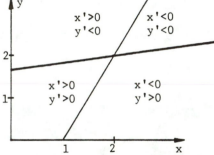

2c. The trajectories will show a similar pattern to those of Problem 1. For large x and y, this system behaves like the example of Eq.(2) and thus we may conclude that $(x,y) \to (4/5,7/5)$ as t → ∞, assuming (x,y) starts in the first quadrant.

3c. For (x,y) above the line $x + y = 2$ we see that $x' < 0$ and thus x must remain bounded. For (x,y) to the right of $x = 1/2$, $y' > 0$ so it appears that y could grow large asymptotic to x = constant. However, this implies a

contradiction (x = constant implies x' = 0, but as y gets

larger, x' gets increasingly negative) and hence we

conclude y must remain bounded and hence $(x,y) \rightarrow$

$(1/2, 3/2)$ as $t \rightarrow \infty$, again assuming they start in the

first quadrant.

4c.   The trajectories will show a similar pattern to those of

Fig. 9.25a.

7a.   Setting the right sides of the equations equal to zero

gives the critical points $(0,0)$, $(0, \epsilon_2/\sigma_2)$, $(\epsilon_1/\sigma_1, 0)$,

and possibly $([\epsilon_1\sigma_2 - \epsilon_2\alpha_1]/[\sigma_1\sigma_2 - \alpha_1\alpha_2], [\epsilon_2\sigma_1 -$

$\epsilon_1\alpha_2]/[\sigma_1\sigma_2 - \alpha_1\alpha_2])$. (The last point can be obtained

from Eq.(22) also). The conditions $\epsilon_2/\alpha_2 > \epsilon_1/\sigma_1$ and $\epsilon_2/$

$\sigma_2 > \epsilon_1/\alpha_1$ imply that $\epsilon_2\sigma_1 - \epsilon_1\alpha_2 > 0$ and $\epsilon_1\sigma_2 - \epsilon_2\alpha_1 < 0$.

Thus either the x coordinate or the y coordinate of

the last critical point is negative so a mixed state is

not possible. The linearized system for $(0,0)$ is x' =

$\epsilon_1 x$ and y'= $\epsilon_2 y$ and thus $(0,0)$ is an unstable equilibrium

point. Similarly, it can be shown [by linearizing the

given system or by using Eq.(21)] that $(0, \epsilon_2/\sigma_2)$ is an

asymptotically stable critical point and that $(\epsilon_1/\sigma_1, 0)$

is an unstable critical point. Thus the fish represented

by y (redear) survive.

7b.   The conditions $\epsilon_1/\sigma_1 > \epsilon_2/\alpha_2$ and $\epsilon_1/\alpha_1 > \epsilon_2/\sigma_2$ imply that

$\varepsilon_2\sigma_1 - \varepsilon_1\alpha_2 < 0$ and $\varepsilon_1\sigma_2 - \varepsilon_2\alpha_1 > 0$ so again one of the coordinates of the fourth point in 7a. is negative and hence a mixed state is not possible. An analysis similar to that in part (a) shows that $(0,0)$ and $(0,\varepsilon_2/\sigma_2)$ are unstable while $(\varepsilon_1/\sigma_1, 0)$ is stable. Hence the bluegill (represented by x) survive in this case.

8b. If B is reduced, it is clear from the answer to part (a) that X is reduced and Y is increased. To determine whether the bluegill will die out, we give an intuitive argument which can be confirmed by doing the analysis. Note that $B/\gamma_1 = \varepsilon_1/\alpha_1 > \varepsilon_2/\sigma_2 = R$ and $R/\gamma_2 = \varepsilon_2/\alpha_2 > \varepsilon_1/\sigma_1 = B$ so that the graph of the lines $1 - x/B - \gamma_1 y/B = 0$ and $1 - y/R - \gamma_2 x/R = 0$ must appear as indicated in the

figure, where critical points are indicated by heavy dots. As B is decreased, X decreases, Y increases (as indicated above) and the point of intersection moves closer to $(0,R)$. If

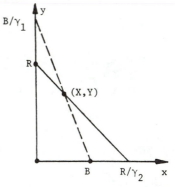

$B/\gamma_1 < R$ coexistence is not possible, and the only critical points are $(0,0)$, $(0,R)$ and $(B,0)$. It can be shown that $(0,0)$ and $(B,0)$ are unstable and $(0,R)$ is asymptotically stable. Hence we conclude, when

coexistence is no longer possible, that $x \to 0$ and $y \to R$ and thus the bluegill population will die out.

11. The presence of a trapping company actually would require a modification of the equations, either by altering the coefficients or by including nonhomogeneous terms on the right sides of the D.E. The effects of indiscriminate trapping could decrease the populations of both rabbits and fox significantly or decrease the fox population which could possibly lead to a large increase in the rabbit population. Over the long run it makes sense for a trapping company to operate in such a way that a consistent supply of pelts is available and to disturb the predator-prey system as little as possible. Thus, the company should trap fox only when their population is increasing, trap rabbits only when their population is increasing, trap rabbits and fox only during the time when both their populations are increasing, and trap neither during the time both their populations are decreasing. In this way the trapping company can have a moderating effect on the population fluctuations, keeping the trajectory close to the center.

13. The critical points of the system are the solutions of the algebraic equations $H(a - \sigma H - \alpha P) = 0$, $P(-c + \gamma H) = 0$. The critical points are $H = 0$, $P = 0$; $H = a/\sigma$, $P = 0$;

and $H = c/\gamma$, $P = a/\alpha - c\sigma/\alpha\gamma = \sigma A/\alpha$ where

$A = a/\sigma - c/\gamma > 0$.

To study the critical point $(0,0)$ we discard the nonlinear terms in the system of D.E. to obtain the corresponding linear system $dH/dt = aH$, $dP/dt = - cP$. The characteristic equation is $r^2 - (a+c)r - ac = 0$ so $r_1 = a, r_2 = - c$. Thus the critical point $(0,0)$ is an unstable saddle point.

To study the critical point $(a/\sigma, 0)$ we let $H = (a/\sigma) + u$, $P = 0 + v$ and substitute in the D.E. to obtain the almost linear system $du/dt = - au - (a\alpha/\sigma)v - \sigma u^2 - \alpha uv$, $dv/dt = \gamma Av + \gamma uv$. The corresponding linear system is $du/dt = - au - (a\alpha/\sigma)v$, $dv/dt = \gamma Av$. The characteristic equation is $r^2 + (a - \gamma A)r - a\gamma A = 0$ so $r_1 = -a, r_2 = \gamma A$. Thus the critical point $(a/\sigma, 0)$ is an unstable saddle point.

To study the critical point $(c/\gamma, \sigma A/\alpha)$ we let $H = (c/\gamma) + u$, $P = (\sigma A/\alpha) + v$ and substitute in the D.E. to obtain the almost linear system $du/dt = - (c\sigma/\gamma)u - (\alpha c/\gamma)v - \sigma u^2 - \alpha uv$, $dv/dt = (\sigma A\gamma/\alpha)u + \gamma uv$. The corresponding linear system is $du/dt = -(c\sigma/\gamma)u - (\alpha c/\gamma)v$, $dv/dt = (\sigma A\gamma/\alpha)u$. The characteristic equation is $r^2 + (c\sigma/\gamma)r + c\sigma A = 0$, so $r_1, r_2 = [- (c\sigma/\gamma) \pm \sqrt{(c\sigma/\gamma)^2 - 4c\sigma A}]/2$. Thus, depending on the sign of the discriminant we have that $(c/\gamma, \sigma A/\alpha)$ is either an asymptotically stable

spiral point or an asymptotically stable improper node.

Thus for nonzero initial data $(H,P) \to (c/\gamma, \sigma A/\alpha)$ as $t \to \infty$.

Section 9.5, Page 509

1a. Assuming that $V(x,y) = ax^2 + cy^2$ we find $V_x(x,y) = 2ax$,

$V_y = 2cy$ and thus Eq.(7) yields $\dot{V}(x,y) = 2ax(-x^3 + xy^2)$

$+ 2cy(-2x^2y - y^3) = -[2ax^4 + 2(2c-a)x^2y^2 + 2cy^4]$. If

we choose a and c to be any positive real numbers with

$2c > a$, then $\dot{V}$ is negative definite. Also, V is

positive definite by Theorem 9.6. Thus by Theorem 9.3

the origin is an asymptotically stable critical point.

1c. Assuming the same form for $V(x,y)$ as in part a, we have

$\dot{V}(x,y) = 2ax(-x^3 + 2y^3) + 2cy(-2xy^2) = -2ax^4 + 4(a-c)xy^3$.

If we choose $a = c > 0$, then $\dot{V}(x,y) = -2ax^4 \le 0$ in any

neighborhood containing the origin and thus $\dot{V}$ is negative

semidefinite and V is positive definite. Theorem 9.3

then concludes that the origin is a stable critical

point. Note that the origin may still be asymptotically

stable, however, the $V(x,y)$ used here is not sufficient

to prove that.

3a. The correct system is $dx/dt = y$ and $dy/dt = -g(x)$.

Since $g(0) = 0$, we conclude that $(0,0)$ is a critical

point.

3b. From the given conditions, the graph of g must be

positive for $0 < x < k$ and negative for $-k < x < 0$. Thus

if $0 < x < k$ then $\int_0^x g(s)ds > 0$,

if $-k < x < 0$ then $\int_0^x g(s)ds = - \int_x^0 g(s)ds > 0$.

Since $V(0,0) = 0$ it follows that $V(x,y) = y^2/2 + \int_0^x g(s)ds$

is positive definite for $-k < x < k$, $-\infty < y < \infty$. Next,

we consider $\dot{V}(x,y) = y[-g(x)] + g(x)y = 0$. Since $\dot{V}(x,y)$

is never positive, we may conclude that it is negative

semidefinite and hence by Theorem 9.3 $(0,0)$ is at least

a stable critical point.

4b. V is positive definite by Theorem 9.6. Since $V_x(x,y) =$

$2x$, $V_y(x,y) = 2y$, we obtain $\dot{V}(x,y) = 2xy - 2y^2 - 2y \sin x$

$= 2y[-y + (x - \sin x)]$. If $x < 0$, then $\dot{V}(x,y) < 0$ for

all $y > 0$. If $x > 0$, choose y so that $0 < y < x - \sin x$.

Then $\dot{V}(x,y) > 0$. Hence V is not a Liapunov function.

4c. Since $V(0,0) = 0$, $1 - \cos x > 0$ for $0 < |x| < 2\pi$ and $y^2 >$

0 for $y \neq 0$, it follows that $V(x,y)$ is positive definite

in a neighborhood of the origin. Next $V_x(x,y) = \sin x$,

$V_y(x,y) = y$, so $\dot{V}(x,y) = (\sin x)(y) + y(-y - \sin x) = -y^2$.

Hence $\dot{V}$ is negative semidefinite and $(0,0)$ is a stable

critical point by Theorem 9.3.

4d. $V(x,y) = (x+y)^2/2 + x^2 + y^2/2 = 3x^2/2 + xy + y^2$ is positive definite by Theorem 9.6. Next $V_x(x,y) = 3x + y$,

$V_y(x,y) = x + 2y$ so $\dot{V}(x,y) = (3x+y)y - (x+2y)(y + \sin x)$

$= 2xy - y^2 - (x+2y) \sin x = 2xy - y^2 - (x+2y)(x - \alpha x^3/6)$

$= -x^2 - y^2 + \alpha(x+2y) x^3/6 = -r^2 + \alpha r^4 (\cos\theta + 2 \sin\theta)\cdot$

$(\cos^3\theta)/6 < -r^2 + r^4/2 = -r^2(1-r^2/2)$. Thus $\dot{V}$ is negative

definite for $r < \sqrt{2}$. From Theorem 9.3 it follows that the

origin is an asymptotically stable critical point.

5.   Let $x = u$ and $y = du/dt$ to obtain the system $dx/dt = y$

and $dy/dt = -c(x)y - g(x)$. Now consider $V(x,y) = y^2/2 +$

$\int_0^x g(s)ds$.

7b.   Since $V_x(x,y) = 2Ax + By$, $V_y(x,y) = Bx + 2Cy$, we have

$\dot{V}(x,y) = (2Ax + By)(ax + by) + (Bx + 2Cy)(cx + dy) = (2Aa$

$+ Bc)x^2 + [2(Ab + Cc) + B(a+d)]xy + (2Cd + Bb)y^2$. We

choose A, B, and C so that $2Aa + Bc = -1$, $2(Ab + Cc) +$

$B(a+d) = 0$, $2Cd + Bb = -1$. The first and third equations

give us A and C in terms of B, respectively. We

substitute in the second equation to find B and then

calculate A and C. The result is given in the text.

7c.   Since $ad - bc > 0$ and $a + d < 0$, we see that $\Delta < 0$ and so

$A > 0$. Using the expressions for A, B, and C found in

part (b) we obtain

$(4AC-B^2)\Delta^2 = [c^2+d^2 + (ad-bc)][a^2+b^2 + (ad-bc)] - (bd+ac)^2$

$$= (a^2+b^2+c^2+d^2)(ad-bc) + (a^2+b^2)(c^2+d^2) + (ad-bc)^2$$
$$- (bd+ac)^2$$
$$= (a^2+b^2+c^2+d^2)(ad-bc) + 2(ad-bc)^2.$$

Since $ad - bc > 0$ it follows that $4AC - B^2 > 0$.

8b. Substituting $x = r \cos\theta$, $y = r \sin\theta$ we find that

$\dot{V}[x(r,\theta), y(r,\theta)] = -r^2 + r(2A \cos\theta + B \sin\theta)F_1[x(r,\theta),$

$y(r,\theta)] + r(B \cos\theta + 2C \sin\theta)G_1[x(r,\theta), y(r,\theta)]$. Now

we make use of the facts that (1) there exists an M such

that $|2A| \leq M$, $|B| \leq M$, and $|2C| \leq M$; and (2) given any

$\varepsilon > 0$ there exists a circle $r = R$ such that $|F_1(x,y)| <$

εr and $|G_1(x,y)| < \varepsilon r$ for $0 \leq r < R$. We have $|2A \cos\theta +$

$B \sin\theta| \leq 2M$ and $|B \cos\theta + 2C \sin\theta| \leq 2M$. Hence

$\dot{V}[x(r,\theta), y(r,\theta)] \leq -r^2 +2Mr(\varepsilon r) +2Mr(\varepsilon r) = -r^2(1 - 4M\varepsilon)$.

If we choose $\varepsilon = M/8$ we obtain $\dot{V}[x(r,\theta), y(r,\theta)] \leq -r^2/2$

for $0 \leq r < R$. Hence $\dot{V}$ is negative definite in $0 \leq r < R$

and from Problem 7c V is positive definite and thus V is

a Liapunov function for the almost linear system.

CHAPTER 10

Section 10.2, Page 521

5. Following the procedures of Eqs.(5) through (8), we set

u(x,y) = X(x)T(t) in the P.D.E. to obtain X"T = 4XT', or

X"/X = 4T'/T, which must be a constant. As shown in the

text this separation constant must be $-\lambda^2$ and thus X" +

λ^2X = 0 and T' + $(\lambda^2/4)$T = 0. Now u(0,t) = X(0) T(t) =

0, for all t > 0, yields X(0) = 0, as discussed after

Eq.(11) and similarly u(2,t) = X(2) T(t) = 0, for all

t > 0, implies X(2) = 0. The D.E. for X has the solution

X(x) = C_1 cos λ x + C_2 sin λ x and X(0) = 0 yields C_1 = 0.

Setting x = 2 in the remaining form of X yields X(2) =

C_2 sin 2λ = 0, which has the solutions 2λ = nπ or λ =

nπ/2, n = 1, 2,... . Note that we exclude n = 0 since

then λ = 0 which would yield X(x) = 0, which is

unacceptable. Hence X(x) = sin (nπx/2), n = 1,2,... .

Finally, the solution of the D.E. for T yields T(t) =

exp($-\lambda^2$t/4) = exp($-n^2\pi^2$t/16). Thus we have found u_n(x,t)

= exp($-n^2\pi^2$t/16) sin (nπx/2). Setting t = 0 in this last

expression indicates that u_n(x,0) has, for the correct

choices of n, the same form as the terms in u(x,0), the

initial condition. Using the principle of superposition

we know that u(x,t) = $c_1 u_1$(x,t) + $c_2 u_2$(x,t) + $c_4 u_4$(x,t) satisfies the P.D.E. and the B.C. and hence we let $t = 0$

to obtain $u(x,0) = c_1 u_1(x,0) + c_2 u_2(x,0) + c_4 u_4(x,0) = c_1 \cdot$

$\sin \pi x/2 + c_2 \sin \pi x + c_4 \sin 2\pi x$. If we choose $c_1 = 2$,

$c_2 = -1$ and $c_4 = 4$ then $u(x,0)$ here will match the

given initial condition, and hence substituting these

values in $u(x,t)$ above then gives the desired solution.

7c.   We seek solutions of the form $u(x,t) = X(x)T(t)$.

Substituting into the P.D.E. yields $X''T + X'T' + XT' =$

$X''T + (X' + X)T' = 0$. Formally dividing by the quantity

$(X' + X)T$ gives the equation $X''/(X' + X) = - T'/T$ in

which the variables are separated. In order for this

equation to be valid on the domain of u it is necessary

that both sides be equal to the same constant λ. Hence

$X''/(X' + X) = - T'/T = \lambda$ or equivalently, $X'' - \lambda(X' + X)$

$= 0$ and $T' + \lambda T = 0$.

7e.   We seek solutions of the form $u(x,y) = X(x)Y(y)$.

Substituting into the P.D.E. yields $X''Y + (x+y)XY'' = X''Y$

$+ xXY'' + yXY'' = 0$. Formally dividing by XY yields $X''/X +$

$xY''/Y + yY''/Y = 0$. From this equation we see that the

presence of the independent variable x multiplying the

term u_{yy} in the original equation leads to the term xY''/Y

when we attempt to separate the variables. It follows

that the argument for a separation constant does not

carry through and we cannot replace the P.D.E. by two

O.D.E.

8. Applying the chain rule to partial differentiation of u
    with respct to x we see that $u_x = u_\xi \xi_x = u_\xi$ $(1/\ell)$ and u_{xx}
    $= u_{\xi\xi}(1/\ell)^2$. Substituting $u_{\xi\xi}/\ell^2$ for u_{xx} in the heat
    equation gives $\alpha^2 u_{\xi\xi}/\ell^2 = u_t$ or $u_{\xi\xi} = (\ell^2/\alpha^2)u_t$. A
    similar argument for the t substitution then yields the
    desired P.D.E.

10. Substituting $u(x,y,t) = X(x)Y(y)T(t)$ in the P.D.E. yields
    $\alpha^2(X''YT + XY''T) = XYT'$, which is equivalent to $X''/X +$
    $Y''/Y = T'/\alpha^2 T$. By keeping the independent variables x
    and y fixed and varying t we see that $T'/\alpha^2 T$ must equal
    some constant σ_1 since the left side of the equation is
    fixed. Hence, $X''/X + Y''/Y = T'/\alpha^2 T = \sigma_1$ or $X''/X = \sigma_1 -$
    Y''/Y and $T' - \sigma_1\alpha^2 T = 0$. By keeping x fixed and varying
    y in the equation involving X and Y we see that $\sigma_1 - Y''/Y$
    must equal some constant σ_2 since the left side of the
    equation is fixed. Hence, $X''/X = \sigma_1 - Y''/Y = \sigma_2$ so $X'' -$
    $\sigma_2 X = 0$ and $Y'' - (\sigma_1 - \sigma_2)Y = 0$. If we assume homogeneous
    B.C. and require that solutions be bounded as $t \rightarrow \infty$, then
    it can be shown by an argument similar to the one in the
    text that σ_1 and σ_2 must be negative constants, $\sigma_1 = -\lambda^2$,
    $\sigma_2 = -\mu^2$. Then the O.D.E. become $X'' + \mu^2 X = 0$, $Y'' +$
    $(\lambda^2 - \mu^2)Y = 0$, and $T' + \alpha^2\lambda^2 T = 0$.

Section 10.3, Page 530

1c. We look for values of T for which sinh 2(x+T) = sinh 2x

for all x. Expanding the left side of this equation

gives sinh 2x cosh 2T + cosh 2x sinh 2T = sinh 2x, which

will be satisfied for all x if we can choose T so that

cosh 2T = 1 and sinh 2T = 0. The only value of T

satisfying these constraints is T = 0. Since T is not

positive we conclude that the function sinh 2x is not

periodic.

1d. We look for values of T for which tanπ(x+T) = tanπx.

Expanding the left side gives tanπ(x+T) = (tanπx +

tanπT)/(1-tanπx tanπT) which is equal to tanπx only for

tanπT = 0. The only positive solutions of this last

equation are T = 1,2,3... and hence tanπx is periodic

with fundamental period T = 1.

2a. First we have $\int_{T}^{a+T} g(x)dx = \int_{0}^{a} g(s)ds$ by letting x = s + T

in the left integral. Now, if $0 \leq a \leq T$, then from

elementary calculus we know that $\int_{a}^{a+T} g(x)dx = \int_{a}^{T} g(x)dx +$

$\int_{T}^{a+T} g(x)dx = \int_{a}^{T} g(x)dx + \int_{0}^{a} g(x)dx$ using the equality

derived above. This last sum is $\int_{0}^{T} g(x)dx$ and thus we

have the desired result.

It is recommended that the given function and its periodic
extension be graphed for several periods for each of the
Problems 5 through 14. In many of these problems it is
necessary to use integration by parts to evaluate the
coefficients, although all the details will be not shown here.

5.   The function represents a sawtooth wave, as indicated in

the figure, and is periodic with period 2ℓ. Thus the

Fourier series is of the form $f(x) = a_0/2 +$

$\sum\limits_{m=1}^{\infty} (a_m \cos m\pi x/\ell + b_m \sin m\pi x/\ell)$ where the coefficients

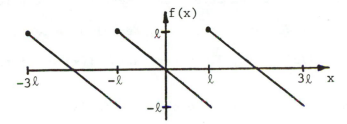

are computed from Eqs. (13) and (14). Substituting for

$f(x)$ in these equations yields $a_0 = (1/\ell) \int\limits_{-\ell}^{\ell} (-x)dx = 0$;

$a_m = (1/\ell) \int\limits_{-\ell}^{\ell} (-x)\cos(m\pi x/\ell)dx = 0$, $m = 1,2...$; and

$b_m = (1/\ell) \int\limits_{-\ell}^{\ell} (-x) \sin(m\pi x/\ell)dx$

$= (x/m\pi)\cos(m\pi x/\ell)\Big|_{-\ell}^{\ell} - (1/m\pi) \int\limits_{-\ell}^{\ell} \cos(m\pi x/\ell)dx$

$= (2\ell \cos m\pi)/m\pi = 2\ell (-1)^m/m\pi.$

Substituting these terms in the above Fourier series for

$f(x)$ yields the desired answer.

9.   The function represents a sawtooth wave, as indicated in

the figure, and although it is periodic with period 2 we

see that the fundamental period is actually 1. Thus the

Fourier series is of the form $f(x) = a_0/2 + \sum\limits_{m=1}^{\infty} (a_m \cos 2m\pi x$

$+ b_m \sin 2m\pi x)$ where the coefficients are computed from

Eqs. (13) and (14) with ℓ replaced by $1/2$. Using the

results of Problem 2c or the discussion just prior to

Example 1 we may change the interval of integration from

$[-1/2, 1/2]$ to $[0,1]$. Substituting $f(x) = x$ into these

equations with the above changes yields $a_0 = 2 \int\limits_0^1 x\, dx =$

$1;\ a_m = 2 \int\limits_0^1 x \cos 2m\pi x\, dx = 0,\ m = 1,2,\ldots;$

$b_m = 2 \int\limits_0^1 x \sin 2m\pi x\, dx = -1/m\pi,\ m = 1,2\ldots\ .$

Substituting these values in the Fourier series for $f(x)$

above gives the desired result. Note that exactly the

same results are obtained if Eqs. (13) and (14) are used

with $\ell = 1$. In this case $b_n = \int\limits_{-1}^0 (x+1)\sin n\pi x\, dx +$

$\int\limits_0^1 x\sin n\pi x\, dx$ which yields $b_n = -2/n\pi$ for n even and b_n

$= 0$ for n odd.

11. In this case $f(x)$ is periodic of period 2π and thus $\ell = \pi$

in Eqs. (9),(13) and (14). The constant a_0 is found to

be $a_0 = (1/\pi) \int_{-\pi}^{0} x\, dx = -\pi/2$ since $f(x)$ is zero on the

interval $[0,\pi]$. Likewise $a_n = (1/\pi) \int_{-\pi}^{0} x \cos nx\, dx =$

$[1 - (-1)^n]/n^2\pi$ using integration by parts. Thus $a_n = 0$

for n even and $a_n = 2/n^2\pi$ for n odd, which may be

written as $a_{2n-1} = 2/(2n-1)^2\pi$ since $2n-1$ is always an

odd number. In a similar fashion $b_n = (1/\pi) \int_{-\pi}^{0} x \sin nx\, dx$

$= (-1)^{n+1}/n$ and thus the desired solution is obtained.

Notice that in this case both cosine and sine terms

appear in the Fourier series for the given $f(x)$.

16. The graph of $f(x)$ is shown in Problem 9. We note that

$f(x)$ is a straight line with a slope of one in any

interval. Thus $f(x)$ has the form x+b in any interval for

the correct value of b. Since $f(x+2) = f(x)$, we may set x

$= -1/2$ to obtain $f(3/2) = f(-1/2)$. Noting that 3/2 is on

the interval $1 < x < 2$ $[f(3/2) = 3/2 + b]$ and that $-1/2$

is on the interval $-1 < x < 0$ $[f(-1/2) = -1/2 + 1]$, we

conclude that $3/2 + b = -1/2 + 1$, or $b = -1$ for the

interval $1 < x < 2$. For the interval $8 < x < 9$ we have

$f(x+8) = f(x+6) = \ldots = f(x)$ by successive applications

of the periodicity condition. Thus for $x = 1/2$ we have

$f(17/2) = f(1/2)$ or $17/2 + b = 1/2$ so $b = -8$ on the interval $8 < x < 9$.

19. Use the procedure outlined in Problem 16 to find $f(x) = x-2$ on the interval $1 < x < 2$. Thus, for instance,

$$b_n = \int_0^2 f(x) \sin n\pi x\, dx = \int_0^1 x \sin n\pi x\, dx + \int_1^2 (x-2) \cdot$$

$\sin n\pi x\, dx = 2(-1)^{n+1}/n\pi$. Similar integrations for a_0 and a_n also yield the same results as in Problem 8.

Section 10.4, Page 538

2. The function to which the series converges is indicated in the figure and is periodic with period 2π. Note that

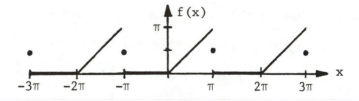

the Fourier series converges to $\pi/2$ at $x = -\pi$, π, etc, even though the function is defined to be zero there. This value $(\pi/2)$ represents the mean value of the left and right hand limits at those points. In $(-\pi, 0)$, $f(x) = 0$ and $f'(x) = 0$ so both f and f' are continuous and have finite limits as $x \to -\pi$ from the right and as $x \to 0$ from the left. In $(0, \pi)$, $f(x) = x$ and $f'(x) = 1$ and again both f and f' are continuous and have limits as $x \to 0$ from the right and as $x \to \pi$ from the left. Since f and f' are piecewise continuous on $[-\pi, \pi)$ the conditions of the Fourier theorem are satisfied.

Substituting for f(x) in Eqs.(2) and (3) with $\ell = \pi$

yields $a_0 = (1/\pi) \int_0^\pi x \, dx = \pi/2$; $a_m =$

$(1/\pi) \int_0^\pi x \cos mx \, dx = (\cos m\pi - 1)/\pi m^2 = 0$

for m even and $= -2/\pi m^2$ for m odd; and $b_m =$

$(1/\pi) \int_0^\pi x \sin mx \, dx = - (\pi \cos m\pi)/m\pi = (-1)^{m+1}/m$, m =

1,2,... . Substituting these values into Eq.(1) with $\ell =$

π yields the desired solution.

3.   The function to which the series converges is shown in

the figure and is periodic with a fundamental period of

π. In $(0,\pi)$, $f(x) = \sin^2 x$ and $f'(x) = 2 \sin x \cos x =$

sin 2x so both f and f' are continuous and have finite

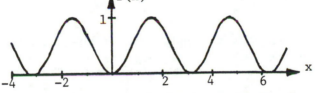

limits as the endpoints of the interval $(0,\pi)$ are

approached from within the interval. Thus f and f' are

piecewise continuous (indeed, they are continuous)

on the interval $(0,\pi)$ and the conditions of the Fourier

theorem are satisfied. Substituting for f(x) in Eqs. (2)

and (3) with $\ell = \pi/2$ and integrating over the interval

$0 \le x < \pi$ (or alternatively, $-\pi/2 \le x < \pi/2$) yields $a_0 =$

$(2/\pi) \int_0^\pi \sin^2 x \, dx = 1$; $a_m = (2/\pi) \int_0^\pi \sin^2 x \cos 2mx \, dx =$

-1/2 if m = 1, otherwise a_m = 0, m = 2,3...; b_m =

$(2/\pi) \int_0^\pi \sin^2 x \sin 2mx \, dx$ = 0, m = 1,2,... . Thus the

Fourier series for f is f(x) = 1/2 - (1/2)cos 2x. This

is not surprising, though, for if we expand $\sin^2 x$ in

terms of the trigonometric half angle formula, we also

have $\sin^2 x$ = 1/2 - (1/2)cos 2x.

6.   The function to which the series converges is shown in

the figure and is periodic of fundamental period 2.   In

[-1,1] f(x) and f'(x) = -2x are both continuous and have

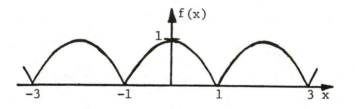

finite limits as the endpoints of the interval are

approached from within the interval. Thus the conditions

of the Fourier Theorem are satisfied. Note that in this

case f(x) and its periodic extension are continuous,

while f'(x) and its periodic extension are only piecewise

continuous. Substituting for f(x) in Eqs.(2) and (3),

with ℓ = 1 yields $a_0 = \int_{-1}^{1} (1-x^2)dx$ = 4/3; a_n =

$\int_{-1}^{1} (1-x^2)\cos n\pi x \, dx = (2/n\pi) \int_{-1}^{1} x \sin n\pi x \, dx$ =

$(-2/n^2\pi^2) [x \cos n\pi x \Big|_{-1}^{1} - \int_{-1}^{1} \cos n\pi x \, dx]$ =

$4(-1)^{n+1}/n^2\pi^2$; $b_n = \int_{-1}^{1} (1-x^2)\sin n\pi x \, dx = 0$.

Substituting these values into Eq.(1) gives the desired series.

10. The solution to the corresponding homogeneous equation is found by methods presented in Section 3.5.1 and is $y(t) = c_1 \cos \omega t + c_2 \sin \omega t$. For the non homogeneous terms we use the method of superposition and consider the sequence of equations $y_n'' + \omega^2 y_n = b_n \sin nt$ for $n = 1,2,3...$. If $\omega > 0$ is not equal to an integer, then the particular solution to this last equation has the form $y_{np} = a_n \cos nt + d_n \sin nt$, as previously discussed in Section 3.6.1. Substituting this form for y_{np} into the equation and solving, we find $a_n = 0$ and $d_n = b_n/(\omega^2 - n^2)$. Thus the formal general solution of the original nonhomogeneous D.E. is $y(t) = c_1 \cos \omega t + c_2 \sin \omega t + \sum_{n=1}^{\infty} b_n (\sin nt)/(\omega^2 - n^2)$, where we have superimposed all the y_{np} terms to obtain the infinite sum. To evaluate c_1 and c_2 we set $t = 0$ to obtain $y(0) = c_1 = 0$ and $y'(0) = \omega c_2 + \sum_{n=1}^{\infty} n \, b_n/(\omega^2 - n^2) = 0$ where we have formally differentiated the infinite series term by term and evaluated it at $t = 0$. (Differentiation of a Fourier Series has not been justified yet and thus we can only

consider this method a formal solution at this point).

Thus $c_1 = 0$, $c_2 = -(1/\omega) \sum\limits_{n=1}^{\infty} n\, b_n/(\omega^2 - n^2)$, which when

substituted into the above series yields the desired

solution.

If $\omega = m$, a positive integer, then the particular

solution of $y_m'' + m^2 y_m = b_m \sin mt$ has the form $y_{mp} =$

$t(a_m \cos mt + d_m \sin mt)$ since $\sin mt$ is a solution of

the related homogeneous D.E.   Substituting y_{mp} into the

D.E. yields $a_m = -b_m/2m$ and $d_m = 0$ and thus the general

solution of the D.E. (when $\omega = m$) is now $y(t) = c_1 \cos mt$

$+ c_2 \sin mt - b_m t(\cos mt)/2m + \sum\limits_{\substack{n=1 \\ n \neq m}}^{\infty} b_n(\sin nt)/(\omega^2 - n^2)$.

To   evaluate c_1 and c_2 we set $y(0) = c_1 = 0$ and $y'(0) =$

$c_2 m - b_m/2m + \sum\limits_{\substack{n=1 \\ n \neq m}}^{\infty} b_n\, n/(\omega^2 - n^2) = 0$.   Thus $c_1 = 0$, $c_2 =$

$b_m/2m^2 - \sum\limits_{\substack{n=1 \\ n \neq m}}^{\infty} b_n\, n/m(\omega^2 - n^2)$, which when substituted

into the equation for $y(t)$ yields the desired solution.

11.   In order to use the results of Problem 10, we must first

find the Fourier series for the given $f(t)$.   Using

Eqs.(2) and (3) with $\ell = \pi$, we find that $a_0 = (1/\pi) \int\limits_0^{\pi} dx$

$- (1/\pi) \int\limits_{\pi}^{2\pi} dx = 0$; $a_n = (1/\pi) \int\limits_0^{\pi} \cos nx\, dx - (1/\pi) \cdot$

$\int\limits_{\pi}^{2\pi} \cos nx\, dx = 0$; and $b_n = (1/\pi) \int\limits_0^{\pi} \sin nx\, dx - (1/\pi) \cdot$

$\int\limits_{\pi}^{2\pi} \sin nx \, dx = 0$ for n even and $= 4/n\pi$ for n odd.   Thus

$f(t) = (4/\pi) \sum\limits_{n=1}^{\infty} \sin(2n-1)t/(2n-1)$.   Comparing this to the

forcing function of Problem 10 we see that b_n of Problem

10 has the specific values $b_{2n} = 0$ and $b_{2n-1} = (4/\pi)/$

$(2n-1)$ in this example. Substituting these into the

answer to Problem 10 yields the desired solution.   Note

that we have assumed ω is not a positive integer.   Note

also, that if the solution to Problem 10 is not

available, the procedure for solving Problem 11 would be

exactly the same as shown in Problem 10.

12.   In this case the Fourier series for f(t) is given by f(t)

$= 1/2 + (4/\pi^2) \sum\limits_{n=1}^{\infty} \cos(2n-1)\pi t/(2n-1)^2$ and thus we may

not use the form of the answer in Problem 10.   The

procedure outlined there, however, is applicable and will

yield the desired solution.

13b. Note that f and f' are continuous at all points where f"

is continuous. Let x_1, ..., x_m be the points in $(-\ell, \ell)$

where f" is not continuous. By splitting up the interval

of integration at these points, and integrating Eq.(3) by

parts twice, we obtain

$$n^2 b_n = \frac{n}{\pi} \sum_{i=1}^{m} [f(x_i+) - f(x_i-)] \cos\frac{n\pi x_i}{\ell} - \frac{n}{\pi}[f(\ell-) - f(-\ell+)] \cos n\pi$$

$$-\frac{\ell}{\pi^2}\sum_{i=1}^{m}[f'(x_i+)-f'(x_i-)]\sin\frac{n\pi x_i}{\ell} - \frac{\ell}{\pi^2}\int_{-\ell}^{\ell}f''(x)\sin\frac{n\pi x}{\ell}\,dx,$$

where we have used the fact that cosine is
continuous. We want the first two terms on the right
side to be zero, for otherwise they grow in magnitude
with n. Hence f must be continuous throughout the closed
interval $[-\ell, \ell]$. The last two terms are bounded, by the
hypotheses on f' and f". Hence $n^2 b_n$ is bounded;
similarly $n^2 a_n$ is bounded. Convergence of the Fourier
series then follows by comparison with $\sum_{n=1}^{\infty} n^{-2}$.

Section 10.5, Page 546

1k. Sketch the graph to assist in answering this problem.

2d. Since sine is an odd function and cosine is an even
function, Property 3 tells us that the product is an odd
function. Hence we use Property 5 to conclude that
$\int_{-\ell}^{\ell}\sin(n\pi x/\ell)\cos(n\pi x/\ell)dx = 0.$

All the functions and their derivatives in Problems 4 through
14 are piecewise continuous on the given intervals and their
extensions. Thus the Fourier Theorem applies in all cases.

4. For the cosine series we use the even extension of the
function given in Eq.(13) and hence

$$f(x) = \begin{cases} 0 & -2 \leq x < -1 \\ 1+x & -1 \leq x < 0 \end{cases} \quad \text{on the interval } -2 \leq x < 0.$$

However, we don't really need this, as the coefficients

in this case are given by Eqs.(7), which just use the
original values for $f(x)$ on $0 < x \leq 2$. Applying Eqs.(7)
we have $\ell = 2$ and thus $a_0 =$

$$(2/2) \int_0^1 (1-x)dx + (2/2) \int_1^2 0\ dx = 1/2.$$ Similarly, $a_n =$

$$(2/2) \int_0^1 (1-x) \cos(n\pi x/2)dx = 4[1-\cos(n\pi/2)]/n^2\pi^2$$ and
$b_n = 0$. Substituting these values in the Fourier series
yields the desired results.

4.   For the sine series, we use Eqs.(8) with $\ell = 2$.
      Thus $a_n = 0$ and $b_n = (2/2) \int_0^1 (1-x) \sin(n\pi x/2)dx = 4[n\pi/2$
      $- \sin(n\pi/2)]/n^2\pi^2$.

5.   The graph of the function to which the series converges
      is shown in the figure. Using Eqs.(7) with $\ell = 2$ we have
      $a_0 = \int_0^1 dx = 1$ and $a_n = \int_0^1 \cos(n\pi x/2)dx = 2\sin(n\pi/2)/n\pi$.

      Thus $a_n = 0$ for n even, $a_n = 2/n\pi$ for $n = 1,5,9,\ldots$ and
      $a_n = -2/n\pi$ for $n = 3,7,11,\ldots$. Hence we may write a_{2n}
      $= 0$ and $a_{2n-1} = 2(-1)^{n+1}/(2n-1)\pi$, which when substituted
      into the series gives the desired answer.

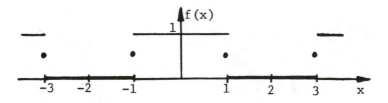

8.   The graph of the function to which the series converges

is indicated in the figure.

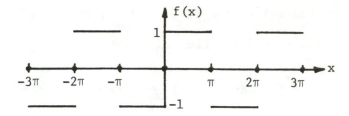

Since we want a sine series, we use Eqs.(8) to find, with

$\ell = \pi$, that $b_n = (2/\pi) \int_0^\pi \sin nx \, dx = 2[1-(-1)^n]/n\pi$ and

thus $b_n = 0$ for n even and $b_n = 4/n\pi$ for n odd.

10. The graph of the function to which the series converges

is shown in the figure.

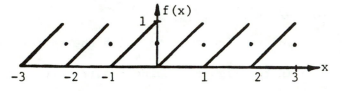

We note that f(x) is specified over its entire

fundamental period (T = 1) and hence we cannot extend f

to make it either an odd or an even function. Using

Eqs.(2) and (3) from Section 10.4 we have ($\ell = 1/2$) $a_0 =$

$2 \int_0^1 x \, dx = 1$, $a_n = 2 \int_0^1 x \cos(2n\pi x)dx = 0$ and $b_n =$

$2 \int_0^1 x \sin(2n\pi x)dx = -1/n\pi$. Substituting these values

into Eq.(1) of Section 10.4 yields the desired solution.

It can also be observed from the above graph that g(x) =

$f(x) - 1/2$ is an odd function. If Eqs.(8) are used with

$g(x)$, then it is found that $g(x) = (-1/\pi) \sum\limits_{n=1}^{\infty} \sin(2n\pi x)/n$

and thus we obtain the same series for $f(x)$ as found

above.

15. Writing $\int\limits_{-\ell}^{\ell} f(x)dx = \int\limits_{-\ell}^{0} f(x)dx + \int\limits_{0}^{\ell} f(x)dx$, substituting x

$= -y$ into the first integral on the right, and using the

fact that $f(x) = -f(-x)$ yields the desired result.

16. To prove property 2 let f_1 and f_2 be odd functions and

let $f(x) = f_1(x) \pm f_2(x)$. Then $f(-x) = f_1(-x) \pm f_2(-x) =$

$- f_1(x) \pm [-f_2(x)] = - f_1(x) \mp f_2(x) = - f(x)$,   so $f(x)$

is odd. Now let $g(x) = f_1(x) f_2(x)$, then $g(-x) =$

$f_1(-x)f_2(-x) = [- f_1(x)][- f_2(x)] = f_1(x)f_2(x) = g(x)$ and

thus $g(x)$ is even. Finally, let $h(x) = f_1(x)/f_2(x)$ and

hence $h(-x) = f_1(-x)/f_2(-x) = [- f_1(x)]/[- f_2(x)]$ =

$f_1(x)/f_2(x) = h(x)$, which says $h(x)$ is also even.

Property 3 is proved in a similar manner.

18. Since $F(x) = \int\limits_{0}^{x} f(t)dt$ we have $F(-x) = \int\limits_{0}^{-x} f(t)dt =$

$- \int\limits_{0}^{x} f(-s)ds$ by letting $t = -s$. If f is an even function,

$f(-s) = f(s)$, we then have $F(-x) = - \int\limits_{0}^{x} f(s)ds = -F(x)$

from the original definition of F. Thus $F(x)$ is an odd

function. The argument is similar if f is odd.

19. Set $x = \ell/2$ in Eq.(6) of Section 10.4. Since we know f is

continuous at this point, we may conclude, by the Fourier
theorem, that the series will converge to $f(\ell/2) = \ell$ at
this point. Thus we have $\ell = \ell/2 + (2\ell/\pi) \cdot$
$\sum\limits_{n=1}^{\infty} (-1)^{n+1}/(2n-1)$, since $\sin[(2n-1)\pi/2] = (-1)^{n+1}$.

21a. Multiplying both sides of the equation by $f(x)$ and
integrating from 0 to ℓ gives $\int_0^\ell [f(x)]^2 dx = \int_0^\ell [f(x) \cdot$
$\sum\limits_{n=1}^{\infty} b_n \sin(n\pi x/\ell)]dx = \sum\limits_{n=1}^{\infty} b_n \int_0^\ell f(x)\sin(n\pi x/\ell)dx = (\ell/2) \cdot$
$\sum\limits_{n=1}^{\infty} b_n^2$. This result is identical to that of Problem 13
of Section 10.4 if we set $a_n = 0$, $n = 0,1,2,\ldots$. In a
similar manner it can be shown that $(2/\ell) \int_0^\ell [f(x)]^2 dx =$
$a_0^2/2 + \sum\limits_{n=1}^{\infty} a_n^2$.

21b. Since $f(x) = x$ and $b_n = 2\ell(-1)^{n+1}/n\pi$, we obtain
$(2/\ell) \int_0^\ell [f(x)]^2 dx = (2/\ell) \int_0^\ell x^2 dx = 2\ell^2/3 = \sum\limits_{n=1}^{\infty} b_n^2 =$
$\sum\limits_{n=1}^{\infty} 4\ell^2/n^2\pi^2 = 4\ell^2/\pi^2 \sum\limits_{n=1}^{\infty} (1/n^2)$ or $\pi^2/6 = \sum\limits_{n=1}^{\infty} (1/n^2)$.

22. We assume that the extensions of f and f' are piecewise
continuous on $[-2\ell, 2\ell]$. Since f is an odd periodic
function of fundamental period 4ℓ it follows from
properties 2 and 3 that $f(x)\cos(n\pi x/2\ell)$ is odd and $f(x) \cdot$
$\sin(n\pi x/2\ell)$ is even. Thus the Fourier coefficients of f
are given by Eqs.(8) with ℓ replaced by 2ℓ; that is, $a_n =$

$$0, \ n = 0,1,2,\ldots \ \text{and} \ b_n = (2/2\ell) \int_0^{2\ell} f(x)\sin(n\pi x/2\ell)dx,$$

$n = 1,2,\ldots$. The Fourier sine series for f is f(x) =

$$\sum_{n=1}^{\infty} b_n \sin(n\pi x/2\ell).$$

23.   From Problem 22 we have

$$b_n = (1/\ell) \int_0^{2\ell} f(x) \sin(n\pi x/2\ell)dx$$

$$= (1/\ell) \int_0^{\ell} f(x) \sin(n\pi x/2\ell)dx + (1/\ell) \int_\ell^{2\ell} f(2\ell-x) \sin(n\pi x/2\ell)dx$$

$$= (1/\ell) \int_0^{\ell} f(x) \sin(n\pi x/2\ell)dx + (1/\ell) \int_\ell^{0} f(s) \sin[n\pi(2\ell-s)/2\ell]ds$$

$$= (1/\ell) \int_0^{\ell} f(x)\sin(n\pi x/2\ell)dx - (1/\ell) \int_0^{\ell} f(s)\cos(n\pi)\sin(n\pi s/2\ell)ds$$

and thus $b_n = 0$ for n even and $b_n =$

$(2/\ell) \int_0^{\ell} f(x) \sin(n\pi x/2\ell)dx$ for n odd. The Fourier

series for f is given in Problem 22, where the b_n are

given above.

Section 10.6, Page 556

1b.   Since the B.C. for this heat conduction problem are

u(0,t) = u(100,t) = 0, t > 0, the solution u(x,t) is

given by Eq.(4) with α^2 = 1.14cm^2/sec, ℓ = 100 cm, and

the coefficients b_n determined by Eq.(6) with the I.C.

$u(x,0) = f(x) = x, \ 0 \leq x \leq 50; \ 100 - x, \ 50 \leq x \leq 100.$

Thus $b_n = (1/50)[\int_0^{50} x \sin(n\pi x/100)dx + \int_{50}^{100} (100-x)$

$\cdot \sin(n\pi x/100)dx] = [400 \sin(n\pi/2)]/n^2\pi^2$, $n = 1,2,\ldots$.

It follows that $u(x,t) = (400/\pi^2) \sum_{n=1}^{\infty} \sin(n\pi/2) \cdot$

$\exp[-n^2\pi^2\alpha^2 t/10000]\sin(n\pi x/100)/n^2$.

2. Since the B.C. for this heat conduction problem are

   $u(0,t) = u(20,t) = 0$, $t > 0$, the solution $u(x,t)$ is given

   by Eq.(4) with $\ell = 20$ cm, and the coefficients b_n

   determined by Eq.(6) with the I.C. $u(x,0) = f(x) = 100^{\circ}C$.

   Thus $b_n = (1/10) \int_0^{20} (100)\sin(n\pi x/20)dx = - 200[(-1)^n$

   $- 1]/n\pi$ and hence $b_{2n} = 0$ and $b_{2n-1} = 400/(2n-1)\pi$.

   Substituting these values into Eq.(4) yields the desired

   solution. For aluminum, we have $\alpha^2 = .86$ cm^2/sec (from

   Table 10.1, Sect.10.1) and thus $u(10,30) =$

   $(400/\pi)\{\exp[-\pi^2(.86)30/400] - (1/3)\exp[-9\pi^2(.86)30/400]\}$

   $= 67.2^{\circ}C$.

3b. Using only one term in the series for $u(x,t)$, we must

    solve the equation $25 = (400/\pi)\exp[-\pi^2(.86)t/400]$ for t.

    Taking the logarithm of both sides and solving for t

    yields $t = 400\ln(16/\pi)/\pi^2(.86) = 76.7$ sec.

4a. Since the B.C. are not homogeneous, we must first find

    the steady state solution. Using Eqs.(12) and (13) we

have v" = 0 with v(0) = 0 and v(ℓ) = 60, which has the

solution v(x) = 60x/ℓ. Thus the transient solution

w(x,t) satisfies the equations $\alpha^2 w_{xx} = w_t$, w(0,t) = 0,

w(ℓ,t) = 0 and w(x,0) = 25 - 60x/ℓ, which are obtained

from Eqs.(16) through (18). The solution of this problem

is given by Eq.(4) with the b_n given by Eq.(6):

$$b_n = (2/\ell) \int_0^\ell (25 - 60x/\ell)\sin(n\pi x/\ell)dx =$$

$[50 + 70(-1)^n]/n\pi$. Finally, u(x,t) = v(x) + w(x,t),

which gives the desired solution when all the terms are

substituted into the indicated equations.

4e. A detailed analysis explaining the stated observation is

given in the paper mentioned in the footnote. The

analysis is based upon the fact that c_n has distinctly

different values depending on whether n is even or odd.

That is $c_n = -20/n\pi$ for n odd and $c_n = 120/n\pi$ for n even

and thus

$$u(x,t) = 60x/\ell + (60/\pi) \sum_{n=1}^\infty (1/n)\exp[-.86(2n)^2\pi^2 t/\ell^2]\sin(n\pi x/\ell)$$

$$- (20/\pi) \sum_{n=1}^\infty [1/(2n-1)]\exp[-.86(2n-1)^2\pi^2 t/\ell^2]\sin[(2n-1)\pi x/\ell].$$

It is then shown that the first sum along with the steady

state solution (a temperature redistribution problem)

converges faster than the second sum, which represents a

heat change. The negative sign on this last sum then

indicates that the steady state temperature is approached from below.

6. Since the B.C. are $u_x(0,t) = u_x(\ell,t) = 0$, $t > 0$, the solution $u(x,t)$ is given by Eq.(32) with the coefficients c_n determined by Eq.(34). Substituting the I.C. $u(x,0) = f(x) = \sin(\pi x/\ell)$ into Eq.(34) yields $c_0 = (2/\ell) \int_0^\ell \sin(\pi x/\ell)dx = 4/\pi$ and $c_n = (2/\ell) \int_0^\ell \sin(\pi x/\ell)$

$\cdot \cos(n\pi x/\ell)dx = (1/\ell) \int_0^\ell \{\sin[(n+1)\pi x/\ell] -$

$\sin[(n-1)\pi x/\ell]\}dx = (1/\pi)\{[1 - \cos(n+1)\pi]/(n+1) - [1 - \cos(n-1)\pi]/(n-1)\} = 0$, n odd; $- 4/(n^2-1)\pi$, n even. Thus $u(x,t) = 2/\pi - (4/\pi) \sum_{n=1}^{\infty} \exp[-4n^2\pi^2\alpha^2 t/\ell^2]\cos(2n\pi x/\ell)$

$/(4n^2-1)$ where we are now summing over even terms by setting n = 2n. As $t \to \infty$ we see that all terms in the series decay to zero except the constant term, $2/\pi$. Hence $\lim_{t\to\infty} u(x,t) = 2/\pi$.

8. The steady-state temperature distribution $v(x)$ must satisfy Eq.(12) and also satisfy the B.C. $v_x(0) = 0$, $v(\ell) = 0$. The general solution $v'' = 0$ is $v(x) = Ax + B$. The B.C. imply that A = 0 and B = 0 so the steady-state solution is $v(x) = 0$.

10a. Substituting $u(x,t) = X(x)T(t)$ into Eq.(1) leads to the two O.D.E. $X'' - \sigma X = 0$ and $T' - \alpha^2\sigma T = 0$. An argument similar to the one in the text implies that we must have

$X(0) = 0$ and $X'(\ell) = 0$. Also, by assuming σ is real and

considering the three cases $\sigma < 0$, $\sigma = 0$, and $\sigma > 0$ we

can show that only the case $\sigma < 0$ leads to nontrivial

solutions of $X'' - \sigma X = 0$ with $X(0) = 0$ and $X'(\ell) = 0$.

Setting $\sigma = -\lambda^2$, we obtain $X(x) = k_1 \sin \lambda x + k_2 \cos \lambda x$.

Now, $X(0) = 0 \to k_2 = 0$ and thus $X(x) = k_1 \sin \lambda x$.

Differentiating and setting $x = \ell$ yields $\lambda k_1 \cos \lambda \ell = 0$.

Since $\lambda = 0$ and $k_1 = 0$ lead to $u(x,t) = 0$, we must choose

λ so that $\cos \lambda \ell = 0$, or $\lambda = (2n-1)\pi/2\ell$, $n = 1,2,3,\ldots$.

These values for λ imply that $\sigma = -(2n-1)^2\pi^2/4\ell^2$ so the

solutions $T(t)$ of $T' - \alpha^2 \sigma T = 0$ are proportional to

$\exp[-(2n-1)^2\pi^2\alpha^2 t/4\ell^2]$. Combining the above results leads

to the desired set of fundamental solutions.

10b. In order to satisfy the I.C. $u(x,0) = f(x)$ we assume

that $u(x,t)$ has the form $u(x,t) =$

$\sum\limits_{n=1}^{\infty} c_n \exp[-(2n-1)^2\pi^2\alpha^2 t/4\ell^2]\sin[(2n-1)\pi x/2\ell]$.   The

coefficients c_n are determined by the requirement that

$u(x,0) = \sum\limits_{n=1}^{\infty} c_n \sin[(2n-1)\pi x/2\ell] = f(x)$. Referring to

Problem 23 of Section 10.5 reveals that such a

representation for $f(x)$ is possible if we choose the

coefficients $c_n = (2/\ell) \int\limits_{0}^{\ell} f(x)\sin[(2n-1)\pi x/2\ell]dx$.

11. To reduce the problem to one with homogeneous B.C. we

first find the steady state solution by solving $v'' = 0$,

subject to the B.C. $v(0) = T$, $v'(\ell) = 0$. We find that

$v(x) = T$. As in the text we write $u(x,t) = v(x) + w(x,t)$

and find that w must satisfy the equation $\alpha^2 w_{xx} = w_t$, the

B.C. $w(0,t) = w_x(\ell,t) = 0$, and the I.C. $w(x,0) = u(x,0) -$

$v(x) = f(x) - T$. Then a formal series expansion for

$w(x,t)$, and hence $u(x,t)$, can be obtained from Problem 10

with $f(x)$ replaced by $f(x) - T$ in determining the Fourier

coefficients.

13. We must solve $v_1''(x) = 0$, $0 \le x \le a$ and $v_2''(x) = 0$,

    $a \le x \le \ell$ subject to the B.C. $v_1(0) = 0$, $v_2(\ell) = T$ and

    the continuity conditions at $x = a$: $v_1(a) = v_2(a)$ and

    $- \kappa_1 A_1 v_1'(a) = -\kappa_2 A_2 v_2'(a)$. The general solutions to the

    two O.D.E. are $v_1(x) = C_1 x + D_1$ and $v_2(x) = C_2 x +$

    D_2. By applying the boundary and continuity conditions

    we may solve for C_1, D_1, C_2 and D_2 to obtain the desired

    solution.

Section 10.7, Page 567

1. Since the initial velocity is zero, the solution is given

   by Eq.(22) with the coefficients k_n given by Eq.(20).

   Substituting $f(x)$ into Eq.(20) yields $k_n =$

   $$(2/\ell)[\int_0^{\ell/2} Ax \sin(n\pi x/\ell)dx + \int_{\ell/2}^{\ell} A(\ell-x)\sin(n\pi x/\ell)dx] =$$

   $[4A\ell\sin(n\pi/2)]/n^2\pi^2$. The desired solution is then obtained by substituting these values into Eq.(22).

3. The solution is given by Eq.(18) with the coefficients k_n and c_n given by Eqs.(20) and (29). Substituting $u(x,0) = f(x) = 0$ into Eq.(20) yields $k_n = 0$, $n = 1,2,\ldots$. From Eq.(29) we see that $c_n = (2/n\pi a) \int_0^\ell g(x)\sin(n\pi x/\ell)dx$, $n = 1,2,\ldots$.

4. Use the results of Problem 3.

6. Assuming that $u(x,t) = X(x)T(t)$ and substituting for u in Eq.(1) leads to the pair of O.D.E. $X'' - \sigma X = 0$, $T'' - a^2\sigma T = 0$. Applying the B.C. $u(0,t) = 0$ and $u_x(\ell,t) = 0$ to $u(x,t)$ we see that we must have $X(0) = 0$ and $X'(\ell) = 0$. By considering the three cases $\sigma < 0$, $\sigma = 0$, and $\sigma > 0$ it can be shown that nontrivial solutions of the problem $X'' - \sigma X = 0$, $X(0) = 0$, $X'(\ell) = 0$ are possible if and only if $\sigma = -(n-1/2)^2\pi^2/\ell^2 = -(2n-1)^2\pi^2/4\ell^2$, $n = 1,2,\ldots$ and the corresponding solutions for $X(x)$ are proportional to $\sin[(2n-1)\pi x/2\ell]$. Using these values for σ we find that $T(t)$ is a linear combination of $\sin[(2n-1)\pi at/2\ell]$ and $\cos[(2n-1)\pi at/2\ell]$. Now, the I.C. $u_t(x,0)$ implies that $T'(0) = 0$ and thus functions of the form $u_n(x,t) = \sin[(2n-1)\pi x/2\ell]\cos[(2n-1)\pi at/2\ell]$, $n = 1,2,\ldots$ satisfy the P.D.E. (1), the B.C. $u(0,t) = 0$, $u_x(\ell,t) = 0$, and the I.C. $u_t(x,0) = 0$. We now seek a superposition of these fundamental solutions u_n that also satisfies the I.C.

$u(x,0) = f(x)$. Thus we assume that $u(x,t) =$

$\sum_{n=1}^{\infty} c_n \sin[(2n-1)\pi x/2\ell]\sin[(2n-1)\pi at/2\ell]$. The I.C. now

implies that we must have $f(x) = \sum_{n=1}^{\infty} c_n \sin[(2n-1)\pi x/2\ell]$.

From Problem 23 of Section 10.5 we see that $f(x)$ can be

represented by such a series and that $c_n =$

$(2/\ell) \int_{0}^{\ell} f(x)\sin[(2n-1)\pi x/2\ell]dx$, $n = 1,2,\ldots$.

Substituting these values into the above series for

$u(x,t)$ yields the desired solution.

8.   Using the chain rule we obtain $u_x = u_\xi \xi_x + u_\eta \eta_x = u_\xi + u_\eta$

since $\xi_x = \eta_x = 1$. Differentiating a second time gives

$u_{xx} = u_{\xi\xi} + 2u_{\xi\eta} + u_{\eta\eta}$. In a similar way we obtain $u_t =$

$u_\xi \xi_t + u_\eta \eta_t = - au_\xi + au_\eta$, since $\xi_t = - a$, $\eta_t = a$. Thus

$u_{tt} = a^2(u_{\xi\xi} - 2u_{\xi\eta} + u_{\eta\eta})$. Substituting for u_{xx} and u_{tt}

in the wave equation, we obtain $u_{\xi\eta} = 0$. Integrating

both sides of $u_{\xi\eta} = 0$ with respect to η yields $u_\xi(\xi,\eta) =$

$\gamma(\xi)$ where γ is an arbitrary function of ξ. Integrating

both sides of $u_\xi(\xi,\eta) = \gamma(\xi)$ with respect to ξ yields

$u(\xi,\eta) = \int \gamma(\xi)d\xi + \psi(\eta) = \phi(\xi) + \psi(\eta)$ where $\psi(\eta)$ is an

arbitrary function of η and $\int \gamma(\xi)d\xi$ is some function of

ξ denoted by $\phi(\xi)$. Thus $u(x,t) = u(\xi(x,t), \eta(x,t)) =$

$\phi(x - at) + \psi(x + at)$.

9.   The graph of $y = \sin(x-at)$ for the various values of t is

indicated in the figure below. Note that the graph of y

= sin x is displaced to the right by the distance   at for

each value of t.

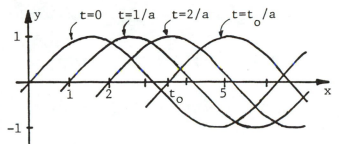

Similarly, the graph of y = ϕ(x + at) would be displaced

to the left by a distance   at   for each t.   Thus

ϕ(x + at) represents a wave moving to the left.

11. Write the equation as $a^2 u_{xx} = u_{tt} + \alpha^2 u$ and assume u(x,t)

= X(x)T(t). The separation constant $\sigma = -\lambda^2$ using the

same arguments as in the text and earlier problems.

12b. Using the hint and the first equation obtained in part

(a) leads to ϕ(x) + ψ(x) = 2ϕ(x) + c = f(x) so ϕ(x) =

(1/2)f(x) − c/2 and ψ(x) = (1/2)f(x) + c/2.   Hence u(x,t)

= ϕ(x − at) + ψ(x + at) = (1/2)[f(x − at) − c] +

(1/2)[f(x + at) + c] = (1/2)[f(x − at) + f(x + at)].

12c. Substituting x + at for x in f(x) yields

$$f(x + at) = \begin{cases} 2 & -1 < x + at < 1 \\ 0 & \text{otherwise} \end{cases}.$$

Subtracting   at   from both sides of the inequality then

yields

$$f(x + at) = \begin{cases} 2 & -1 -at < x < 1 - at \\ \\ 0 & \text{otherwise} \end{cases}.$$

13b. From part (a) we have $\psi(x) = -\phi(x)$ and hence

$$\psi(x) = (1/2a) \int_{x_0}^{x} g(\xi)d\xi - \phi(x_0).$$

13c. $u(x,t) = \phi(x-at) + \psi(x+at) = -(1/2a) \int_{x_0}^{x-at} g(\xi)d\xi + \phi(x_0)$

$$+ (1/2a) \int_{x_0}^{x+at} g(\xi)d\xi - \phi(x_0) = (1/2a)[\int_{x_0}^{x+at} g(\xi)d\xi -$$

$$\int_{x_0}^{x-at} g(\xi)d\xi] = (1/2a)[\int_{x_0}^{x+at} g(\xi)d\xi + \int_{x-at}^{x_0} g(\xi)d\xi] =$$

$$(1/2a) \int_{x-at}^{x+at} g(\xi)d\xi.$$

14. Use the result of Problem 5.

17. Substituting $u(r,\theta,t) = R(r)\Theta(\theta)T(t)$ into the P.D.E.

yields $R''\Theta T + R'\Theta T/r + R\Theta''T/r^2 = R\Theta T''/a^2$ or equivalently

$R''/R + R'/rR + \Theta''/r^2 = T''/a^2 T$. In order for this equation

to be valid for $0 < r < r_0$, $0 \le \theta \le 2\pi$, $t > 0$, it

is necessary that both sides of the equation be equal to

the same constant σ. Otherwise, by keeping r and θ fixed

and varying t, one side would remain unchanged while the

other side varied. Thus we arrive at the two equations

$T'' - \sigma a^2 T = 0$ and $r^2 R''/R + rR'/R - \sigma r^2 = -\Theta''/\Theta$. By an

argument similar to the one above we conclude that both

sides of the last equation must be equal to the same

constant δ. This leads to the two equations $r^2R" + rR' -$

$(\sigma r^2 + \delta)R = 0$ and $\Theta" + \delta\Theta = 0$. Since the circular

membrane is continuous, we must have $\Theta(2\pi) = \Theta(0)$, which

requires $\delta = \mu^2$, μ a non-negative integer. The condition

$\Theta(2\pi) = \Theta(0)$ is also known as the periodicity condition.

Since we also desire solutions which vary periodically in

time, it is clear that the separation constant σ should

be negative, $\sigma = -\lambda^2$. Thus we arrive at the three

equations $r^2R" + rR' + (\lambda^2r^2-\mu^2)R = 0$, $\Theta" + \mu^2\Theta = 0$, and

$T" + \lambda^2a^2T = 0$.

Section 10.8, Page 577

1a. Assuming that $u(x,y) = X(x)Y(y)$ leads to the two O.D.E.

$X" - \sigma X = 0$, $Y" + \sigma Y = 0$. The B.C. $u(0,y) = 0$, $u(a,y) =$

0 imply that $X(0) = 0$ and $X(a) = 0$. Thus nontrivial

solutions to $X" - \sigma X = 0$ which satisfy these boundary

conditions are possible only if $\sigma = - (n\pi/a)^2$, $n =$

$1,2,\ldots$; the corresponding solutions for $X(x)$ are

proportional to $\sin(n\pi x/a)$. The B.C. $u(x,0) = 0$ implies

that $Y(0) = 0$. Solving $Y" - (n\pi/a)^2Y = 0$ subject to this

condition we find that $Y(y)$ must be proportional to

$\sinh n\pi y/a$. The fundamental solutions $u_n(x,y) =$

$\sin(n\pi x/a)\sinh(n\pi y/a)$, $n = 1,2,\ldots$ satisfy Laplace's

equation and the homogeneous B.C. We assume that $u(x,y)$

$= \sum\limits_{n=1}^{\infty} c_n \sin(n\pi x/a)\sinh(n\pi y/a)$, where the coefficients

c_n are determined from the B.C. $u(x,b) = g(x) =$

$\sum\limits_{n=1}^{\infty} c_n \sin(n\pi x/a)\sinh(n\pi b/a)$. It follows that

$c_n \sinh(n\pi b/a) = (2/a) \int\limits_{0}^{a} g(x)\sin(n\pi x/a)dx$, $n = 1,2,\ldots$.

1b. Substituting for $g(x)$ in the equation for c_n we have

$c_n \sinh(n\pi b/a) = (2/a)[\int\limits_{0}^{a/2} x \sin(n\pi x/a)dx +$

$\int\limits_{a/2}^{a} (a-x)\sin(n\pi x/a)dx] = [4a \sin(n\pi/2)]/n^2\pi^2$, $n = 1,2,\ldots,$

so $c_n = [4a \sin(n\pi/2)]/[n^2\pi^2 \sinh(n\pi b/a)]$. Substituting

these values for c_n in the above series yields the

desired solution.

2. In solving the D.E. $Y'' - \lambda^2 Y = 0$, one normally writes

$Y(y) = c_1 \sinh \lambda y + c_2\cosh \lambda y$. However, since we have $Y(b)$

$= 0$, it is advantageous to rewrite Y as $Y(y) =$

$d_1 \sinh \lambda (b-y) + d_2\cosh \lambda (b-y)$, where d_1, d_2 are also

arbitrary constants and can be related to c_1, c_2 using

the appropriate hyperbolic trigonometric identities. The

important thing, however, is that the second form also

satisfies the D.E. and thus $Y(y) = d_1 \sinh\lambda(b-y)$ satisfies

the D.E. and the homogeneous B.C. $Y(b) = 0$. The rest of

the problem follows the pattern of Problem 1.

4. Following the pattern of Problem 3, one would consider

adding the solutions of four problems, each with only one

non-homogeneous B.C. It is also possible to consider

adding the solutions of only two problems, each with only

two non-homogeneous B.C., as long as they involve the

same variable. For instance, one such problem would be

$u_{xx} + u_{yy} = 0$, $u(x,0) = 0$, $u(x,b) = 0$, $u(0,y) = k(y)$,

$u(a,y) = f(y)$, which has the fundamental solutions

$u_n(x,y) = [c_n \sinh(n\pi x/b) + d_n \cosh(n\pi x/b)]\sin(n\pi y/b)$.

Assuming $u(x,y) = \sum\limits_{n=1}^{\infty} u_n(x,y)$ and using the B.C. $u(0,y) =$

$k(y)$ we obtain $d_n = (2/a) \int\limits_{0}^{a} k(y)\sin(n\pi y/b)dy$. Using the

B.C. $u(a,y) = f(y)$ we obtain $c_n \sinh(n\pi a/b) +$

$d_n \cosh(n\pi a/b) = (2/a) \int\limits_{0}^{a} f(y)\sin(n\pi y/b)dy$, which can be

solved for c_n, since d_n is already known.

5.   Using Eq.(17) and following the same arguments as

presented in the text, we find that $R(r) = k_1 r^n + k_2 r^{-n}$

and $\Theta(\theta) = c_1 \sin n\theta + c_2 \cos n\theta$, for n a positive

integer, and $u_0(r,\theta) = 1$ for $n = 0$. Since we require

that $u(r,\theta)$ be bounded as $r \to \infty$, we conclude that $k_1 = 0$.

The fundamental solutions are therefore $u_n(r,\theta) =$

$r^{-n}\cos n\theta$, $v_n(r,\theta) = r^{-n}\sin n\theta$, $n = 1,2,...$ together with

$u_0(r,\theta) = 1$. Assuming that u can be expressed as a

linear combination of the fundamental solutions we have

$u(r,\theta) = c_0/2 + \sum_{n=1}^{\infty} r^{-n}(c_n \cos n\theta + k_n \sin n\theta)$. The B.C.

requires that $u(a,\theta) = c_0/2 + \sum_{n=1}^{\infty} a^{-n}(c_n \cos n\theta +$

$k_n \sin n\theta) = f(\theta)$ for $0 \leq \theta < 2\pi$. This is precisely the

Fourier series representation for $f(\theta)$ of period 2π and

thus $a^{-n} c_n = (1/\pi) \int_0^2 f(\theta)\cos n\theta d\theta$, $n = 0,1,2,\ldots$ and

$a^{-n} k_n = (1/\pi) \int_0^2 f(\theta)\sin n\theta d\theta$, $n = 1,2,\ldots$.

7.   Again we let $u(r,\theta) = R(r)\Theta(\theta)$ and thus we have $r^2 R''$

+ $rR' - \sigma R = 0$ and $\Theta'' + \sigma\Theta = 0$, with $R(0)$ bounded and

the B.C. $\Theta(0) = \Theta(\alpha) = 0$. Since $u(r,\theta)$ must be single

valued we conclude that $\sigma = \lambda^2$ (λ^2 real) and thus $\Theta(\theta) =$

$c_1 \cos \lambda\theta + c_2 \sin \lambda\theta$. The B.C. $\Theta(0) = 0 \to c_1 = 0$ and

the B.C. $\Theta(\alpha) = 0 \to \lambda = n\pi/\alpha$, $n = 1,2,\ldots$. Substituting

these values into Eq.(28) we obtain $R(r) = k_1 r^{n\pi/\alpha} +$

$k_2 r^{-n\pi/\alpha}$. However $k_2 = 0$ since $R(0)$ must be bounded,

and thus the fundamental solutions are $u_n(r,\theta) =$

$r^{n\pi/\alpha} \sin(n\pi\theta/\alpha)$. The desired solution may now be formed

using previously discussed procedures.

8.   Since neither $\sinh y$ nor $\cosh y$ are bounded as $y \to \infty$, we

must write the solution to $Y'' - (n\pi/a)^2 Y = 0$ as $Y(y) =$

$c_1 \exp[n\pi y/a] + c_2 \exp[-n\pi y/a]$. Thus we must choose $c_1 = 0$

so that $u(x,y) = X(x)Y(y) \to 0$ as $y \to \infty$.

13.   Assuming that $u(x,y) = X(x)Y(y)$ and substituting into

Eq.(1) leads to the two O.D.E. $X" - \sigma X = 0$, $Y" + \sigma Y = 0$.

The B.C. $u(x,0) = 0$, $u_y(x,b) = 0$ imply that $Y(0) = 0$ and

$Y'(b) = 0$. For nontrivial solutions to exist for $Y" + \sigma Y$

$= 0$ with these B.C. we find that σ must take the values

$(2n-1)^2\pi^2/b^2$, $n = 1,2,\ldots$; the corresponding solutions

for $Y(y)$ are proportional to $\sin[(2n-1)\pi y/b]$. Solutions

to $X" - [(2n-1)^2\pi^2/b^2]X = 0$ are of the form $X(x) =$

$A \sinh[(2n-1)\pi x/2b] + B \cosh[(2n-1)\pi x/2b]$. However, the

boundary condition $u(0,y) = 0$ implies that $X(0) = B = 0$.

It follows that the fundamental solutions are $u_n(x,y) =$

$c_n \sinh[(2n-1)\pi x/2b]\sin[(2n-1)\pi y/2b]$, $n = 1,2,\ldots$. To

satisfy the remaining B.C. at $x = a$ we assume that we can

represent the solution $u(x,y)$ in the form $u(x,y) =$

$\sum\limits_{n=1}^{\infty} c_n \sinh[(2n-1)\pi x/2b]\sin[(2n-1)\pi y/2b]$. The

coefficients c_n are determined by the B.C. $u(a,y) =$

$\sum\limits_{n=1}^{\infty} c_n \sinh[(2n-1)\pi a/2b]\sin[(2n-1)\pi y/2b] = f(y)$. By

properly extending f as a periodic function of period 4b

as in Problem 23, Section 10.5, we find that the

coefficients c_n are given by $c_n \sinh[(2n-1)\pi a/2b] =$

$(2/b) \int\limits_{0}^{b} f(y)\sin[(2n-1)\pi y/2b]dy$, $n = 1,2,\ldots$.

CHAPTER 11

Section 11.1, Page 592

1b. Since the B.C. at $x = 1$ is nonhomogeneous, the B.V.P. is nonhomogeneous.

1d. The D.E. may be written $y" + (\lambda-x^2)y = 0$ and is thus homogeneous, as are both B.C.

1e. Since the D.E. contains the nonhomogeneous term 1, the B.V.P. is nonhomogeneous.

2b. Eq.(iii) is linear and separable. Using the latter approach we have $d\mu/\mu = [(Q-P')/P]dx$ and thus $\ln \mu = \int^x [Q(s)/P(s)]ds - \ln P$. Taking the exponential of both sides yields Eq.(iv). The choice of x_0 simply alters the constant of integration, which is immaterial here.

3b. Since $P(x) = x^2$ and $Q(x) = x$, we find that $\mu(x) = (1/x^2)\exp[\int^x_{x_0} (s/s^2)ds] = k/x$, where k is an arbitrary constant which may be set equal to 1. It follows that Bessel's equation takes the form $(xy')' + (x-\nu^2/x)y = 0$.

Section 11.2, Page 599

1. If $\lambda < 0$, the general solution of the D.E. is $y = c_1\sinh\sqrt{\mu}x + c_2 \cosh\sqrt{\mu}x$ where $-\lambda = \mu$. The two B.C. require that $c_2 = 0$ and $c_1 = 0$ so $\lambda < 0$ is not an

eigenvalue. If $\lambda = 0$, the general solution of the D.E. is

$y = c_1 + c_2 x$. The B.C. require that $c_1 = 0$, $c_2 = 0$ so

$\lambda = 0$ is not an eigenvalue.

If $\lambda > 0$, the general solution of the D.E. is $y =$

$c_1 \sin \sqrt{\lambda} x + c_2 \cos \sqrt{\lambda} x$. The B.C. require that $c_2 = 0$,

$c_1 \cos\sqrt{\lambda} = 0$. The second condition is satisfied for $c_1 \neq$

0 if $\lambda = [(2n-1)\pi/2]^2$, $n = 1,2,\ldots$. Thus the eigenvalues

are $\lambda_n = [(2n-1)\pi/2]^2$, $n = 1,2,\ldots$ with corresponding

eigenfunctions $\phi_n(x) = \sin[(2n-1)\pi x/2]$, $n = 1,2,\ldots$.

4. Note that the problem is identical to Problem 1 with λ

    replaced by $-\lambda$.

5. If $\lambda = 0$, the general solution of the D.E. is $y = c_1 +$

    $c_2 x$. The B.C. $y(0) + y'(0) = 0$ requires $c_1 + c_2 = 0$ and

    the B.C. $y(1) = 0$ requires $c_1 + c_2 = 0$ and thus $\lambda = 0$ is

    an eigenvalue with corresponding eigenfunction $\phi_0(x) =$

    $1-x$.

    If $\lambda < 0$, set $-\lambda = \mu^2$ to obtain $y = c_1 \cos \mu x + c_2 \sin \mu x$.

    In this case the B.C. require $c_1 + \mu c_2 = 0$ and $c_1 \cos \mu$

    $+ c_2 \sin \mu = 0$ which yields nontrivial solutions for c_1

    and c_2 (i.e., $c_1 = -\mu c_2$) if and only if $\tan \mu = \mu$. By

    plotting on the same graph $f(\mu) = \mu$ and $g(\mu) = \tan \mu$, we

    see that they intersect at $\mu_0 = 0$ ($\mu = 0 \to \lambda = 0$, which

    has already been discussed), $\mu_1 \approx 4.493$ (which is just

    to the left of the vertical asymptote of $\tan \mu$ at $\mu =$

$3\pi/2$), and for larger values $\mu_n \simeq (2n+1)\pi/2$. Since

$\lambda_n = -\mu_n^2$, we have $\lambda_1 \simeq -20.2$, $\lambda_n \simeq -(2n+1)^2\pi^2/4$ and

$\phi_n = \sin \mu_n x - \mu_n \cos \mu_n x$.

If $\lambda > 0$, the general solution of the D.E. is $y(x) =$

$c_1 \cosh\sqrt{\lambda}x + c_2 \sinh\sqrt{\lambda}x$. The B.C. respectively

require that $c_1 + \sqrt{\lambda}c_2 = 0$ and $c_1 \cosh \sqrt{\lambda} + c_2 \sinh \sqrt{\lambda} =$

0 and thus λ must satisfy $\tanh \sqrt{\lambda} = \sqrt{\lambda}$ in order to have

nontrivial solutions. The only solution of this equation

is $\lambda = 0$ and thus there are no positive eigenvalues.

8.   If $\lambda > 0$, the general solution of the D.E. is $y =$

$c_1 \sin\sqrt{\lambda}x + c_2 \cos\sqrt{\lambda}x$. The B.C. require that $c_2 - \sqrt{\lambda} c_1 =$

0, $(\sin \sqrt{\lambda} + \sqrt{\lambda} \cos\sqrt{\lambda})c_1 + (\cos \sqrt{\lambda} - \sqrt{\lambda} \sin\sqrt{\lambda})c_2 = 0$.

In order to have nontrivial solutions λ must satisfy

$(\lambda-1)\sin \sqrt{\lambda} - 2\sqrt{\lambda}\cos \sqrt{\lambda} = 0$. In this case $c_2 = \sqrt{\lambda}c_1$ and

thus $\phi_n = \sin \sqrt{\lambda_n}x + \sqrt{\lambda_n} \cos \sqrt{\lambda_n}x$. If $\lambda \neq 1$, the

eigenvalue equation is equivalent to $\tan \sqrt{\lambda} = 2\sqrt{\lambda}/(\lambda-1)$

and thus by graphing $f(\sqrt{\lambda}) = \tan\sqrt{\lambda}$ and $g(\sqrt{\lambda}) =$

$2\sqrt{\lambda}/(\lambda-1)$ we can estimate the eigenvalues. Since $g(\sqrt{\lambda})$

has a vertical asymptote at $\lambda = 1$ and $f(\sqrt{\lambda})$ has a

vertical asmymptote at $\sqrt{\lambda} = \pi/2$, we see that $1 < \sqrt{\lambda_1} <$

$\pi/2$. By interating numerically, we find $\sqrt{\lambda_1} \simeq 1.3$ and

thus $\lambda_1 \simeq 1.69$. For large values of n, we see from the

graph that $\sqrt{\lambda_n} \simeq (n-1)\pi$, which are the zeros of $\tan \sqrt{\lambda}$.

Thus $\lambda_n \simeq (n-1)^2\pi^2$ for large n.

For $\lambda \leq 0$, the discussion follows the pattern of earlier problems.

10a. Assuming $y = s(x)u$, we have $y' = s'u + su'$ and $y'' = s''u + 2s'u' + su''$ and thus the D.E. becomes $su'' + (2s'+4s)u' + [s'' + 4s' + (4+9\lambda)s]u = 0$. Setting $2s' + 4s = 0$ we find $s(x) = e^{-2x}$ and the D.E. becomes $u'' + 9\lambda u = 0$. The B.C. $y(0) = 0$ yields $s(0)u(0) = 0$, or $u(0) = 0$ since $s(0) \neq 0$. The B.C. at ℓ is $y'(\ell) = s'(\ell)u(\ell) + s(\ell)u'(\ell) = e^{-2\ell}(-2u(\ell) + u'(\ell)) = 0$ and thus $u'(\ell) - 2u(\ell) = 0$. Thus the B.V.P. satisfied by $u(x)$ is $u'' + 9\lambda u = 0$, $u(0) = 0$, $u'(\ell) - 2u(\ell) = 0$.

If $\lambda < 0$, the general solution of the D.E. $u'' + 9\lambda u = 0$ is $u = c_1 \sinh 3\mu x + c_2 \cosh 3\mu x$ where $-\lambda = \mu^2$. The B.C. require that $c_2 = 0$, $c_1(3\mu \cosh 3\mu\ell - 2 \sinh 3\mu\ell) = 0$. In order to have nontrivial solutions μ must satisfy the equation $3\mu/2 = \tanh 3\mu\ell$. A graphical analysis reveals that for $\ell \leq 1/2$ this equation has no solutions for $\mu \neq 0$ so there are no negative eigenvalues for $\ell \leq 1/2$. If $\ell > 1/2$ there is one solution and hence one negative eigenvalue with eigenfunction $\phi_{-1}(x) = e^{-2x} \sinh 3\mu x$.

If $\lambda = 0$, the general solution of the D.E. $u'' + 9\lambda u = 0$ is $u = c_1 + c_2 x$. The B.C. require that $c_1 = 0$, $c_2(1-2\ell) = 0$ so nontrivial solutions are possible only if $\ell = 1/2$. In this case the eigenfunction is $\phi_0(x) = x e^{-2x}$.

If $\lambda > 0$, the general solution of the D.E. $u'' + 9\lambda u = 0$
is $u = c_1 \sin 3\sqrt{\lambda}x + c_2 \cos 3\sqrt{\lambda}x$. The B.C. require
that $c_2 = 0$, $c_1(3\sqrt{\lambda} \cos 3\sqrt{\lambda}\ell - 2\sin 3\sqrt{\lambda}\ell) = 0$. In
order to have nontrivial solutions λ must satisfy the
equation $\sqrt{\lambda} = (2/3)\tan 3\sqrt{\lambda}\ell$. A graphical analysis
reveals that there is an infinite number of solutions to
this eigenvalue equation. Thus the eigenfunctions are
$\phi_n(x) = e^{-2x} \sin 3\sqrt{\lambda_n}x$ where the eigenvalues λ_n satisfy
$\sqrt{\lambda_n} = (2/3)\tan 3\sqrt{\lambda_n}\ell$.

12. This is an Euler equation. If $\lambda = 1$ the general solution
of the D.E. is $y = c_1 x + c_2 x\ln x$ and the B.C. require that
$c_1 = c_2 = 0$ and thus $\lambda = 1$ is not an eigenvalue. If $\lambda \neq$
1, $y = c_1 x + c_2 x^\lambda$ is the general solution and the B.C.
require that $c_1 + c_2 = 0$ and $2c_1 + c_2 2^\lambda - (c_1 + \lambda c_2 2^{\lambda-1})$
$= 0$. Thus nontrivial solutions exist if and only if $\lambda =$
$2(1-2^{-\lambda})$. The graphs of $f(\lambda) = \lambda$ and $g(\lambda) = 2(1-2^{-\lambda})$
intersect only at $\lambda = 1$ (which has already been
discussed) and $\lambda = 0$. Thus the only eigenvalue is $\lambda = 0$
with corresponding eigenfunction $\phi(x) = x-1$ (since
$c_1 = -c_2$).

14a. For positive λ, the general solution of the D.E. is $y =$
$c_1 \sin\sqrt{\lambda}x + c_2 \cos\sqrt{\lambda}x$. The B.C. require that $\sqrt{\lambda}c_1 +$
$\alpha c_2 = 0$, $c_1 \sin\sqrt{\lambda} + c_2 \cos\sqrt{\lambda} = 0$. Nontrivial solutions
exist if and only if $\sqrt{\lambda}\cos\sqrt{\lambda} - \alpha \sin\sqrt{\lambda} = 0$. If $\alpha = 0$

this equation is satisfied by the sequence $\lambda_n = [(2n-1) \cdot \pi/2]^2$, $n = 1,2,\ldots$. If $\alpha \neq 0$, λ must satisfy the equation $\sqrt{\lambda}/\alpha = \tan\sqrt{\lambda}$. A plot of the graphs of $f(\sqrt{\lambda}) = \sqrt{\lambda}/\alpha$ and $g(\sqrt{\lambda}) = \tan\sqrt{\lambda}$ reveals that there is an infinite sequence of positive eigenvalues for $\alpha < 0$ and $\alpha > 0$.

14b. By procedures shown previously, the cases $\lambda < 0$ and $\lambda = 0$, when $\alpha < 1$, lead to only the trivial solution and thus by part a all real eigenvalues are positive. For $0 < \alpha < 1$, the graph of $f(\sqrt{\lambda})$ and $g(\sqrt{\lambda})$ (see part a) intersect once on $0 < \sqrt{\lambda} < \pi/2$. As α approaches 1 from below, the slope of $f(\sqrt{\lambda})$ decreases and thus the intersection point approaches zero.

15. Using the D.E. for ϕ_m and following the hint yields:

$$\int_0^{\ell} \phi_m'' \phi_n \, dx + \lambda_m \int_0^{\ell} \phi_m \phi_n \, dx = 0.$$

Integrating the first term by parts yields:

$$\phi_m' \phi_n \Big|_0^{\ell} - \int_0^{\ell} \phi_m' \phi_n' \, dx = -\lambda_m \int_0^{\ell} \phi_m \phi_n \, dx.$$

Upon utilizing the B.C. the first term on the left vanishes and thus $\int_0^{\ell} \phi_n' \phi_m' \, dx = \lambda_m \int_0^{\ell} \phi_m \phi_n \, dx.$

Similarly, the D.E. for ϕ_n yields $\int_0^{\ell} \phi_m' \phi_n' \, dx =$

$\lambda_n \int_0^\ell \phi_n \phi_m dx$ and thus $(\lambda_n - \lambda_m) \int_0^\ell \phi_m \phi_n dx = 0$. If

$\lambda_n \neq \lambda_m$ the desired result follows.

16b. The general solution of the D.E. is $y = c_1 \sin \mu x + c_2 \cdot$

cos $\mu x + c_3 \sinh \mu x + c_4 \cosh \mu x$ where $\lambda = \mu^4$. The B.C.

require that $c_2 + c_4 = 0$, $-c_2 + c_4 = 0$, $c_1 \sin\mu\ell + c_2 \cdot$

cos $\mu\ell + c_3 \sinh \mu\ell + c_4 \cosh \mu\ell = 0$, $c_1 \cos \mu\ell - c_2 \cdot$

sin $\mu\ell + c_3 \cosh \mu\ell + c_4 \sinh \mu\ell = 0$. The first two

equations yield $c_2 = c_4 = 0$, and the last two have

nontrivial solutions if and only if $\sin \mu\ell \cosh \mu\ell$ $-$

cos $\mu\ell \sinh \mu\ell = 0$. In this case the third equation yields

$c_3 = -c_1 \sin \mu\ell /\sinh \mu\ell$ and thus the desired

eigenfunctions are obtained.

Section 11.3, Page 612

1a. In Problem 1 of Section 11.2 we found the eigenfunctions

to be $\sin[(2n-1)\pi x/2]$, $n = 1,2,\ldots$ and thus we must

choose k_n so that $\int_0^1 \{k_n \sin[(2n-1)\pi x/2]\}^2 dx = 1$, since

the weight function $r(x) = 1$ [See Eq.(20)]. Evaluating

the integral yields $k_n^2/2 = 1$ and thus $k_n = \sqrt{2}$ and the

desired normalized eigenfunctions are obtained.

1c. Note here that $\phi_0(x) = 1$ satisfies Eq.(34) and hence it

is already normalized.

1e. From Problem 9 of Section 11.2 we have $e^x \sin n\pi x$, n =

     1,2,... as the eigenfunctions and thus k_n must be chosen

     so that $\int_0^1 k_n^2 e^{2x} \sin^2 n\pi x \, dx = 1$ [again r(x) = 1]. Setting

     $\sin^2 n\pi x = (1 - \cos 2n\pi x)/2$ and integrating by parts yields

     $k_n^2(e^2-1)n^2\pi^2/4(n^2\pi^2+1) = 1$. Solving for k_n and

     multiplying the above eigenfunctions yields the desired

     normalized eigenfunctions.

2b. Using Eq.(34) with r(x) = 1, we find that the

     coefficients of the series (32) are determined by a_n =

     $(f,\phi_n) = \sqrt{2} \int_0^1 x \sin[(2n-1)\pi x/2]dx =$

     $(4\sqrt{2}/(2n-1)^2\pi^2)\sin(2n-1)\pi/2$. Thus Eq.(32) yields f(x) =

     $$\frac{4\sqrt{2}}{\pi^2} \sum_{n=1}^{\infty} \frac{(-1)^{n-1}}{(2n-1)^2} \sqrt{2} \sin[(2n-1)\pi x/2], \quad 0 \le x \le 1, \text{ which}$$

     agrees with the expansion using the approach developed in

     Problem 23 of Section 10.5.

3a. In this case $\phi_n(x) = (\sqrt{2}/\alpha_n)\cos\sqrt{\lambda_n}\, x$, where α_n =

     $(1 + \sin^2\sqrt{\lambda_n})^{1/2}$. Thus Eq.(34) yields a_n =

     $(\sqrt{2}/\alpha_n) \int_0^1 \cos\sqrt{\lambda_n}x \, dx = \sqrt{2} \sin\sqrt{\lambda_n}\,/\alpha_n\sqrt{\lambda_n}.$

4a. In this case L[y] = y" + y' + 2y is not of the form shown

     in Eq.(3) and thus the B.V.P. is not self adjoint.

4d. In this case $L[y] = [(1+x^2)y']' + y$ and thus the D.E. has the form shown in Eq.(3). However, the B.C. are not

separated and thus we must determine by integration

whether Eq.(8) is satisifed. Therefore, for u and v

satisfying the B.C., integration by parts yields the

following: $(L[u],v)$

$$= \int_0^1 \{[(1+x^2)u']'+u\} \, v \, dx = vu'(1+x^2)\Big|_0^1 - \int_0^1 \{(1+x^2)v'u'+uv\} \, dx$$

$$= vu'(1+x^2)\Big|_0^1 - uv'(1+x^2)\Big|_0^1 + \int_0^1 \{[(1+x^2)v']' + v\} \, u \, dx$$

$= (u,L[v])$ since the integrated terms add to zero with

the given B.C. Thus the B.V.P. is self-adjoint.

7a. Substituting ϕ for y in the D.E., multiplying both sides

by ϕ, and integrating from 0 to 1 yields $\lambda \int_0^1 r \, \phi^2 \, dx =$

$\int_0^1 \{-[p(x)\phi']' \, \phi + q(x) \, \phi^2\} dx$. Integrating the first

term on the right side once by parts, we obtain

$$\lambda \int_0^1 r \, \phi^2 \, dx = -p(1)\phi'(1)\phi(1) + p(0)\phi'(0)\phi(0) +$$

$\int_0^1 (p\phi'^2 + q\phi^2)dx$. If $a_2 \neq 0$, $b_2 \neq 0$, then $\phi'(1) =$

$-b_1\phi(1)/b_2$ and $\phi'(0) = -a_1\phi(0)/a_2$ and the result

follows. If $a_2 = 0$, then $\phi(0) = 0$ and the boundary term

at 0 will be missing. A similar result is obtained if b_2

$= 0$.

9a. Using $\phi(x) = u(x) + iv(x)$ in Eq.(4) we have $L[\phi] =$

$L[u(x) + iv(x)] = \lambda r(x)[u(x)+iv(x)]$. Using the

linearity of L and the fact that λ and $r(x)$ are real we

have $L[u(x)] + i\ L[v(x)] = \lambda r(x)u(x) + i\ \lambda r(x)v(x)$.

Equating the real and imaginary parts shows that both u

and v satisfy Eq.(1). The B.C. Eq.(2) are also satisfied

by both u and v, using the same arguments, and thus both

u and v are eigenfunctions.

9c. From part b we have $v(x) = cu(x)$ and thus $\phi(x) = u(x) +$

$icu(x) = (1+ic)u(x)$.

10b. See Problem 9 of Section 11.2.

10c. The eigenfunctions are $\phi_n(x) = e^x \sin n\pi x$, $n = 1,2\ldots$ and

the weight function is $r(x) = 1$. Thus $\int_0^1 r(x)\ \phi_n(x) \cdot$

$\phi_m(x)\ dx = \int_0^1 e^{2x}\sin(n\pi x)\sin(m\pi x)dx$, which, by direct

integration, is not necessarily zero when $m \neq n$.

11. If $\lambda = 1$, the general solution to the D.E. is $y = c_1 x +$

$c_2 x \ln x$. The B.C. require that $c_1 = 0$, $2c_1 + 2(\ln 2)c_2$

$= 0$ so $c_1 = c_2 = 0$ and $\lambda = 1$ is not an eigenvalue. If

$\lambda \neq 1$, the general solution to the D.E. is $y = c_1 x +$

$c_2 x^\lambda$. The B.C. require that $c_1 + c_2 = 0$, $2c_1 + 2^\lambda c_2 =$

0. Nontrivial solutions exist if and only if $2^\lambda - 2 = 0$.

If λ is real, this equation has no solution (other than

$\lambda = 1$) and again $y = 0$ is the only solution to the boundary value problem. Suppose that $\lambda = a + bi$ with $b \neq$

0. Then $2^{\lambda} = 2^{a+bi} = 2^{a}2^{bi} = 2^{a} \exp(ib \ln 2)$, which upon

substitution into $2^{\lambda} = 2$ yields the equation $\exp(ib \ln 2)$

$= 2^{1-a}$. It follows that $a = 1$ and $b \ln 2 = 2n\pi$ or $b =$

$2n\pi/\ln 2$, $n = \pm 1, \pm 2, \ldots$. Thus the only eigenvalues

of the problem are $\lambda_{n} = 1 + i(2n\pi/\ln 2)$, $n = \pm 1, \pm 2, \ldots$.

12b. The general solution of the D.E. is $y = c_{1} + c_{2}x +$

$c_{3} \sin\sqrt{\lambda}x + c_{4}\cos\sqrt{\lambda}x$. The B.C. require that $c_{1} + c_{4} = 0$,

$c_{4} = 0$, $c_{1} + c_{2}\ell + c_{3} \sin \sqrt{\lambda}\ell + c_{4} \cos \sqrt{\lambda}\ell = 0$, $c_{2} +$

$\sqrt{\lambda} c_{3} \cos \sqrt{\lambda}\ell - \sqrt{\lambda} c_{4} \sin \sqrt{\lambda}\ell = 0$. Thus $c_{1} = c_{4} = 0$

and for nontrivial solutions to exist λ must satisfy the

equation $\sqrt{\lambda}\ell \cos \sqrt{\lambda}\ell - \sin \sqrt{\lambda}\ell = 0$. In this case $c_{2} =$

$(-\sqrt{\lambda} \cos\sqrt{\lambda}\ell)(c_{3})$ and it follows that the eigenfunction ϕ_{1}

is given by $\phi_{1}(x) = \sin(\sqrt{\lambda_{1}} x) - \sqrt{\lambda_{1}} x \cos \sqrt{\lambda_{1}}\ell$ where λ_{1}

is the smallest positive solution of the equation $\sqrt{\lambda}\ell =$

$\tan \sqrt{\lambda}\ell$. A graphical or numerical estimate of λ_{1}

reveals that $\lambda_{1} \simeq (4.49)^{2}/\ell^{2}$.

12c. The eigenvalue equation is $2(1-\cos x) = x \sin x$, where x

$= \sqrt{\lambda}\ell$. Graphing $f(x) = 2(1-\cos x)$ and $g(x) = x \sin x$ we

see that there is an intersection at $x = 2\pi$. In

addition, it appears there might be an intersection for

$0 < x < 1$. Using a Taylor series representation for

$f(x)$ and $g(x)$ about $x = 0$, however, shows there is no

intersection for $0 < x < 1$. Of course $x = 0$ is also an

intersection, which yields $\lambda = 0$, which gives the trivial

solution and hence $\lambda = 0$ is not an eigenvalue.

Section 11.4, Page 625

la. We must first find the eigenvalues and normalized

eigenfunctions of the associated homogeneous problem $y" +$

$\lambda y = 0$, $y(0) = 0$, $y(1) = 0$. This problem has the

solutions $\phi_n(x) = k_n \sin n\pi x$, for $\lambda_n = n^2\pi^2$, $n = 1,2,\ldots$.

Choosing k_n so that $\int_0^1 \phi_n^2 \, dx = 1$ we find $k_n = \sqrt{2}$. Hence

the solution of the original nonhomogeneous problem is

given by $y = \sum_{n=1}^{\infty} b_n \phi_n(x)$, where the coefficients b_n are

found from Eq.(12), $b_n = c_n/(\lambda_n - 2)$ where c_n is given by

$c_n = \sqrt{2} \int_0^1 x \sin n\pi x \, dx$ (Eq.9). [Note that the original

problem can be written as $-y" = 2y + x$ and therefore

comparison with Eq.(1) yields $f(x) = x$]. Integrating the

expression for c_n by parts yields $c_n = \sqrt{2} (-1)^{n+1}/n\pi$ and

thus $y = \sum_{n=1}^{\infty} \frac{\sqrt{2}(-1)^{n+1}}{(n^2\pi^2 - 2)n\pi} \sqrt{2} \sin n\pi x$.

lb. From Problem la of Section 11.3 we have $\phi_n =$

$\sqrt{2} \sin[(2n-1)\pi x/2]$ for $\lambda_n = (2n-1)^2\pi^2/4$ and from Problem

2b of that section we have $c_n = 4\sqrt{2} (-1)^{n-1}/(2n-1)^2\pi^2$.

Substituting these values into $b_n = c_n/(\lambda_n - 2)$ and $y =$

$\sum\limits_{n=1}^{\infty} b_n \phi_n$ yields the desired result.

1c. Referring to Problem 1c of Section 11.3 we have $y = b_0 +$
$\sum\limits_{n=1}^{\infty} b_n (\sqrt{2} \cos n\pi x)$, where $b_n = c_n/(\lambda_n - 2)$ for $n =$
$0, 1, 2, \ldots$. The rest of the calculations follow those of
Problem 1a.

1e. Note that the associated eigenvalue problem is the same
as for Problem 1a and that $|1-2x| = 1-2x$ for $0 \le x \le 1/2$
while $|1-2x| = 2x-1$ for $1/2 \le x \le 1$.

3a. Since $\mu = \pi^2$ is an eigenvalue of the corresponding
homogeneous equation, Theorem 11.5 tells us that a
solution will exist only if $-(a+x)$ is orthogonal to the
corresponding eigenfunction $\sqrt{2} \sin\pi x$. Thus we require
$\int_0^1 (a+x) \sin\pi x \, dx = 0$, which yields $a = -1/2$. With $a =$
$-1/2$, we find that $y_p = (x-1/2)/\pi^2$ and $y_c = c \sin\pi x +$
$d \cos\pi x$ by methods of Chapter 3. Setting $y = y_c + y_p$ and
choosing d to satisfy the B.C. we obtain the desired
family of solutions.

3b. Note that in this case $\mu = 4\pi^2$ and $\phi_2 = \sqrt{2} \sin 2\pi x$ are
the eigenvalue and eigenfunction respectively of the
corresponding homogeneous equation.

3c. In this case a solution will exist only if $-a$ is

orthogonal to $\sqrt{2} \cos \pi x$., that is if $\int_0^1 a \cos x\pi dx = 0$.

Since this condition is valid, a family of solutions

exists.

4.   Since $\overset{\infty}{\underset{n=1}{\Sigma}} c_n \phi_n(x)$ converges to zero we have: $\overset{\infty}{\underset{n=1}{\Sigma}} c_n \phi_n(x) = $
    0. Multiplying and integrating as suggested yields

$$\int_0^1 [\overset{\infty}{\underset{n=1}{\Sigma}} c_n \phi_n(x)]r(x) \quad \phi_m(x)dx = 0 \text{ or}$$

$$\overset{\infty}{\underset{n=1}{\Sigma}} c_n \int_0^1 r(x)\phi_n(x) \phi_m(x)dx = 0. \text{ The integral that}$$

multiplies c_n is just δ_{nm} [Eq.(22) of Section 11.3].

Thus the infinite sum becomes c_m and the last equation

yields $c_m = 0$.

8.   A twice differentiable function v satisfying the boundary

conditions can be found by assuming that $v = ax + b$.

Thus $v(0) = b = 1$ and $v(1) = a + 1$ while $v'(1) = a$.

Hence $2a + 1 = -2$ or $a = -3/2$ and $v(x) = 1 - 3x/2$.

Assuming $y = u + v$ we have $(u+v)" + 2(u+v) = u" + 2u + $

$2(1-3x/2) = 2 - 4x$ or $u" + 2u = -x$, $u(0) = 0$, $u(1) + $

$u'(1) = 0$ which is the same as Example 1 of the text.

9a.   From Eq.(30) we assume $u(x,t) = \overset{\infty}{\underset{n=1}{\Sigma}} b_n(t) \phi_n(x)$, where the

ϕ_n are the eigenfunctions of the related eigenvalue

problem $y" + \lambda y = 0$, $y(0) = 0$, $y'(1) = 0$ and the $b_n(t)$

are given by Eq.(42). From Problem 1a, Section 11.3 we

have $\phi_n = \sqrt{2} \sin[(2n-1)\pi x/2]$ and $\lambda_n = (2n-1)^2 \pi^2/4$. To evaluate Eq.(42) we need to calculate $\alpha_n = \int_0^1 \sin(\pi x/2) \cdot$

$\sqrt{2} \sin[(2n-1)\pi x/2]dx$ [Eq.(41) with $r(x) = 1$ and $f(x) =$

$\sin(\pi x/2)$], which is zero except for $n = 1$ in which case

$$\alpha_1 = \sqrt{2}/2, \text{ and } \gamma_n = \int_0^1 (-x) \sqrt{2} \sin[(2n-1)\pi x/2]dx$$

[Eq.(35) with $F(x,t) = -x$]. From Problem 1b we have $\gamma_n =$
$-4\sqrt{2}(-1)^{n+1}/(2n-1)^2\pi^2 = -c_n$, $n = 1,2,\ldots$. Setting $\gamma_n =$
$-c_n$ in Eq.(42) we then have

$$b_1 = \frac{\sqrt{2}}{2} e^{-\pi^2 t/4} - c_1 \int_0^t e^{-\pi^2(t-s)/4} ds$$

$$= \frac{\sqrt{2}}{2} e^{-\pi^2 t/4} - \frac{4c_1}{\pi^2} e^{-\pi^2(t-s)/4} \Big|_0^t$$

$$= \frac{\sqrt{2}}{2} e^{-\pi^2 t/4} - \frac{4c_1}{\pi^2} + \frac{4c_1}{\pi^2} e^{-\pi^2 t/4} \quad \text{and}$$

similarly $b_n = -c_n \int_0^t e^{-\lambda_n(t-s)} ds = -(c_n/\lambda_n) e^{-\lambda_n(t-s)} \Big|_0^t =$
$-(c_n/\lambda_n)(1 - e^{-\lambda_n t})$, where $\lambda_n = (2n-1)^2\pi^2/4$, $n =$
$2,3,\ldots$. Substituting these values for b_n along with ϕ_n
$= \sqrt{2} \sin[(2n-1)\pi x/2]$ into the series for $u(x,t)$ yields
the solution to the given problem.

9d. In this case $\alpha_n = 0$ for all n and γ_n is given by $\gamma_n =$

$$\int_0^1 e^{-t}(1-x)\sqrt{2} \sin[(2n-1)\pi x/2]dx = e^{-t} \int_0^1 (1-x) \cdot$$

$\sqrt{2}\sin[(2n-1)\pi x/2]dx$. This last integral can be

written as the sum of two integrals, each of which has

been evaluated in either Problem 2a or 2b of Section

11.3. Letting c_n denote the value obtained, we then have

$$\gamma_n = c_n e^{-t} \text{ and thus } b_n = c_n \int_0^t e^{-\lambda_n(t-s)} e^{-s} ds =$$

$$c_n e^{-\lambda_n t} \int_0^t e^{(\lambda_n-1)s} ds = [c_n/(\lambda_n-1)](e^{-t}-e^{-\lambda_n t}), \text{ where } \lambda_n =$$

$(2n-1)^2\pi^2/4$. Substituting these values into Eq.(30)

yields the desired solution.

11a. Using the approach of Problem 10 we find that $v(x)$

satisfies $v'' = 2$, $v(0) = 1$, $v(1) = 0$. Thus $v(x) = x^2 +$

$c_1 x + c_2$ and the B.C. yield $v(0) = c_2 = 1$ and $v(1) = 1 +$

$c_1 + 1 = 0$ or $c_1 = -2$. Hence $v(x) = x^2 - 2x + 1$ and

$w(x,t) = u(x,t) - v(x)$ where, from Problem 10, we have

$w_t = w_{xx}$, $w(0,t) = 0$, $w(1,t) = 0$ and $w(x,0) = x^2 - 2x +$

$2 - v(x) = 1$. This last problem can be solved by methods

of this section or by methods of Chapter 10. Using the

approach of this section we have $w(x,t) = \sum_{n=1}^{\infty} b_n(t)\phi_n(x)$

where $\phi_n(x) = \sqrt{2} \sin n\pi x$ [which are the normalized

eigenfunctions of the associated eigenvalue problem $y'' +$

$\lambda y = 0$, $y(0) = 0$, $y(1) = 0$] and the b_n are given by

Eq.(42). Since the P.D.E. for $w(x,t)$ is homogeneous

Eq.(42) reduces to $b_n = \alpha_n e^{-\lambda_n t}$ ($\lambda_n = n^2\pi^2$ from the above eigenvalue problem), where $\alpha_n = \int_0^1 1 \cdot \sqrt{2} \sin n\pi x \, dx =$

$\sqrt{2}[1-(-1)^n]/n\pi.$ Thus $u(x,t) = x^2-2x + 1 + \sum_{n=1}^{\infty} \frac{\sqrt{2}[1-(-1)^n]}{n\pi} \cdot$

$e^{-n^2\pi^2 t} \sqrt{2} \sin n\pi x$, which simplifies to the desired

solution.

14a. Using variation of parameters we assume $y_p(x) = u_1(x) +$

$xu_2(x)$. Then $y_p' = u_2$ since we require $u_1' + xu_2' = 0$.

Differentiating again yields $y_p'' = u_2'$ and thus $u_2' = -f(x)$

by substitution into the D.E. Hence $u_2(x) = - \int_0^x f(s)ds,$

$u_1' = xf(x)$, and $u_1(x) = \int_0^x s \, f(s)ds$. Therefore $y_p =$

$\int_0^x s \, f(s)ds - x \int_0^x f(s)ds = - \int_0^x (x-s)f(s)ds$ and

$\phi(x) = c_1 + c_2 x - \int_0^x (x-s)f(s)ds.$

14c. From parts a and b we have

$\phi(x) = x \int_0^1 (1-s) \, f(s) \, ds - \int_0^x (x-s) \, f(s) \, ds$

$= \int_0^x x(1-s)f(s)ds + \int_x^1 x(1-s)f(s)ds - \int_0^x (x-s)f(s)ds$

$= \int_0^x (x-xs-x+s)f(s)ds + \int_x^1 x(1-s)f(s)ds$

$= \int_0^x s(1-x)f(s)ds + \int_x^1 x(1-s)f(s)ds.$

16b. In this case $y_1(x) = \sin x$ and $y_2(x) = \sin(1-x)$ [assume

$y_2(x) = c_1 \cos x + c_2 \sin x$, let $x = 1$, solve for c_2 in

terms of c_1 using $y(1) = 0$ and then let $c_1 = \sin 1$].

Using these functions for y_1 and y_2 we find $W(y_1,y_2) =$

$-\sin 1$ and thus $G(x,s) = -\sin s \sin(1-x)/(-\sin 1)$, since

$p(x) = 1$, for $0 \le s \le x$. A similar calculation verifies

$G(x,s)$ for $x \le s \le 1$.

16c. Since $W(y_1,y_2)(x) = y_1(x)y_2'(x) - y_2(x)y_1'(x)$ we find that

$[p(x)W(y_1,y_2)(x)]' = p'(x)[y_1(x)y_2'(x) - y_2(x)y_1'(x)]$

$+ p(x)[y_1'(x)y_2'(x) + y_1(x)y_2''(x) - y_2'(x)y_1'(x) - y_2(x)y_1''(x)]$

$= y_1[py_2']' - y_2[py_1']' = y_1[q(x)y_2] - y_2]q(x)y_1] = 0.$

16d. Let $c = p(x)W(y_1,y_2)(x)$. If $0 \le s \le x$, then $G(x,s) =$

$-y_1(s)y_2(x)/c$. Since the first argument in $G(s,x)$ is

less than the second argument, the bottom expression of

formula (iv) must be used to determine $G(s,x)$. Thus,

$G(s,x) = -y_1(s)y_2(x)/c$. A similar argument holds if

$x \le s \le 1$.

17c. In general $y(x) = c_1 \cos x + c_2 \sin x$. For $y'(0) = 0$ we

must choose $c_2 = 0$ and thus $y_1(x) = \cos x$. For $y(1) = 0$

we have $c_1\cos 1 + c_2\sin 1 = 0$, which yields $c_2 =$

$-c_1(\cos 1)/\sin 1$ and thus $y_2(x) = c_1\cos x - c_1(\cos 1)\cdot$

$\sin x/ \sin 1 = c_1(\sin 1 \cos x - \cos 1 \sin x)/\sin 1 =$

$\sin(1-x)$ [by setting $c_1 = \sin 1$]. Furthermore, $W(y_1,y_2)$

$= -\cos 1$ and thus

$$G(x,s) = \begin{cases} \dfrac{\cos s \, \sin (1-x)}{\cos 1} & 0 \le s \le x \\[3mm] \dfrac{\cos x \, \sin (1-s)}{\cos 1} & x \le s \le 1 \end{cases},$$

and hence $\phi(x) = \int_0^x [\cos s \, \sin(1-x) \, f(s)/\cos 1]ds +$

$\int_x^1 [\cos x \, \sin(1-s) \, f(s)/\cos 1]ds$ is the solution of the

given B.V.P.

Section 11.5, Page 637

2a. The general solution of the D.E. is $y = c_1 J_0(\sqrt{\lambda}x) +$

$c_2 Y_0(\sqrt{\lambda} \, x)$ by Eq.(7). The B.C. at $x = 0$ requires that c_2

$= 0$, and the B.C. at $x = 1$ requires that $c_1 \sqrt{\lambda} J_0'(\sqrt{\lambda}) =$

0. For $\lambda = 0$ we have $\phi_0(x) = J_0(0) = 1$ and if λ_n is the

n^{th} positive root of $J_0'(\sqrt{\lambda}) = 0$ then $\phi_n(x) = J_0(\sqrt{\lambda_n}x)$.

Note that for $\lambda = 0$ the D.E. becomes $(xy')' = 0$, which

has the general solution $y = c_1 \ln x + c_2$. To satisfy the

bounded conditions at $x = 0$ we must choose $c_1 = 0$, thus

obtaining the same solution as above.

2b. For $n \ne 0$, set $y = J_0(\sqrt{\lambda_n}x)$ in the D.E. and integrate

from 0 to 1 to obtain $-\int_0^1 (xJ_0')'dx = \lambda_n \int_0^1 x \, J_0(\sqrt{\lambda_n}x)dx.$

Integrating the left side of this equation yields

$$\int_0^1 (x J_0')'dx = x \, J_0'(\sqrt{\lambda_n}x) \Big|_0^1 = J_0'(\sqrt{\lambda_n}) - 0 = 0 \text{ since}$$

the λ_n are eigenvalues from part a. Thus

$\int_0^1 x\, J_0(\sqrt{\lambda_n}x)dx = 0.$ For other n and m, we let $L[y] =$

$-(xy')'$. Then $L[J_0(\sqrt{\lambda_n}x)] = \lambda_n x\, J_0(\sqrt{\lambda_n}x)$ and

$L[J_0(\sqrt{\lambda_m}x)] = \lambda_m x\, J_0(\sqrt{\lambda_m}x)$. Multiply the first

equation by $J_0(\sqrt{\lambda_m}x)$, the second by $J_0(\sqrt{\lambda_n}x)$,

subtract the second from the first, and integrate from 0

to 1 to obtain $\int_0^1 \{J_0(\sqrt{\lambda_m}x)L[J_0(\sqrt{\lambda_n}x)] -$

$J_0(\sqrt{\lambda_n}x)L[J_0(\sqrt{\lambda_m}x)]\}dx = (\lambda_n - \lambda_m)\int_0^1 x\, J_0(\sqrt{\lambda_n}x) \cdot$

$J_0(\sqrt{\lambda_m}x)dx$. Again the left side is zero after each term

is integrated by parts once, as was done above. If

$\lambda_n \neq \lambda_m$, the result follows with $\phi_n(x) = J_0(\sqrt{\lambda_n}x)$.

2c. We assume that $y = b_0 + \sum_{n=1}^{\infty} b_n\, J_0(\sqrt{\lambda_n}x)$. Since

$-[xJ_0'(\sqrt{\lambda_n}x)]' = \lambda_n x\, J_0(\sqrt{\lambda_n}x),$ n = 0,1,..., we find

that $-(xy')' = x \sum_{n=1}^{\infty} \lambda_n\, b_n\, J_0(\sqrt{\lambda_n}x)$ [note that $\lambda_0 = 0$

and hence b_0 is missing on the right]. Now assume $f(x)/x$

$= c_0 + \sum_{n=1}^{\infty} c_n\, J_0(\sqrt{\lambda_n}x)$. Multiplying both sides by

$xJ_0(\sqrt{\lambda_m}x)$, integrating from 0 to 1 and using the

orthogonality relations of part b, we find $c_n =$

$\int_0^1 f(x)J_0(\sqrt{\lambda_n}x)dx / \int_0^1 x\, J_0^2(\sqrt{\lambda_n}x)dx,$ n = 0,1,2,... .

[Note that $c_0 = 2\int_0^1 f(x)dx$ since the denominator can be integrated.] Substituting the series for y and $f(x)/x$

into the D.E., using the above result for $-(xy')'$ and

simplifying we find that $(\mu b_0 + c_0) + \sum\limits_{n=1}^{\infty} [c_n - b_n(\lambda_n - \mu)] \cdot$

$J_0(\sqrt{\lambda_n} x) = 0$. Thus $b_0 = -c_0/\mu$ and $b_n = c_n/(\lambda_n - \mu)$, n =

1,2,..., where $\sqrt{\lambda_n}$ are obtained from $J_0'(\sqrt{\lambda_n}) = 0$.

4a. Let $L[y] = -[(1 - x^2)y']'$. Then $L[\phi_n] = \lambda_n \phi_n$ and $L[\phi_m] =$

$\lambda_m \phi_m$. Multiply the first equation by ϕ_m, the second by

ϕ_n, subtract the second from the first, and integrate

from 0 to 1 to obtain $\int\limits_{0}^{1} (\phi_m L[\phi_n] - \phi_n L[\phi_m])dx = (\lambda_n - \lambda_m) \cdot$

$\int\limits_{0}^{1} \phi_n \phi_m \, dx$. The integral on the left side can be shown

to be 0 by integrating each term once by parts. Since

$\lambda_n \neq \lambda_m$ if $m \neq n$, the result follows. Note that the

result may also be written as $\int\limits_{0}^{1} P_{2m-1}(x) \, P_{2n-1}(x)dx = 0$,

$m \neq n$.

4b. First let $f(x) = \sum\limits_{n=1}^{\infty} c_n \phi_n(x)$, multiply both sides by $\phi_m(x)$,

and integrate term by term from $x = 0$ to $x = 1$. The

orthogonality condition yields $c_n = \int\limits_{0}^{1} f(x) \, \phi_n(x)dx/$

$\int\limits_{0}^{1} \phi_n^2(x)dx$, n = 1,2,... and it is understood that $\phi_n(x)$

$= P_{2n-1}(x)$. Now assume $y = \sum\limits_{n=1}^{\infty} b_n \phi_n(x)$. As in Problem 2 and in the text $-[(1-x^2)y']' = \sum_{n=1}^{\infty} \lambda_n b_n \phi_n$ since the ϕ_n are

eigenfunctions. Thus substitution of the series for y and

f into the D.E. and simplification yields

$\sum_{n=1}^{\infty} [b_n(\lambda_n - \mu) - c_n] \phi_n(x) = 0.$ Hence $b_n = c_n/(\lambda_n - \mu),$

n = 1,2,... and the desired solution is obtained [after

setting $\phi_n(x) = P_{2n-1}(x)$].

Section 11.6, Page 643

1a. Since $u(x,0) = 0$ we have $Y(0) = 0.$ However, since the

other two boundaries are given by $y = 2x$ and $y = 2(x-2)$

we cannot separate x and y dependence and thus neither X

nor Y satisfy homogeneous B.C. at both end points.

2.   This problem is very similar to the example worked in the

text. The fundamental solutions satisfying the P.D.E.

(3), the B.C. $u(1,t) = 0,$ $t \geq 0$ and the finiteness

condition are given by Eqs. (15) and (16). Thus assume

u(r,t) is of the form given by Eq. (17). The I.C.

require that $u(r,0) = \sum_{n=1}^{\infty} c_n J_0(\lambda_n r) = 0$ and $u_t(r,0) =$

$\sum_{n=1}^{\infty} \lambda_n a k_n J_0(\lambda_n r) = g(r).$ From Eq.(23) of Section 11.5 we

obtain $c_n = 0$ and

$$\lambda_n k_n a = \int_0^1 rg(r)J_0(\lambda_n r)dr / \int_0^1 rJ_0^2(\lambda_n r)dr, \quad n = 1,2,\ldots \,.$$

4.   This problem is the same as Problem 17 of Section 10.7.

The periodicity condition requires that μ of that problem

be an integer and thus substituting $\mu^2 = n^2$ into the

previous results yields the given equations.

5a. Substituting $u(r,\theta,z) = R(r)\Theta(\theta)Z(z)$ into Laplace's

equation yields $R''\Theta Z + R'\Theta Z/r + R\Theta''Z/r^2 + R\Theta Z'' = 0$ or

equivalently $R''/R + R'/rR + \Theta''/r^2\Theta = - Z''/Z = \sigma$. In

order to satisfy arbitrary B.C. it can be shown that σ

must be negative, so assume $\sigma = - \lambda^2$, and thus $Z'' - \lambda^2 Z =$

0 and, after some algebra, it follows that $r^2 R''/R + rR'/R$

$+ \lambda^2 r^2 = - \Theta''/\Theta = \alpha$. The periodicity condition $\Theta(0) =$

$\Theta(2\pi)$ requires that $\sqrt{\alpha}$ be an integer n so $\alpha = n^2$. Thus

$r^2 R'' + rR' + (\lambda^2 r^2 - n^2)R = 0$, $\Theta'' + n^2\Theta = 0$, and

$Z'' - \lambda^2 Z = 0$.

5b. If $u(r,\theta,z)$ is independent of θ, then the $\Theta''/r^2\Theta$ term

does not appear in the second equation of part a and thus

$R''/R + R'/rR = - Z''/Z = - \lambda^2$, from which the desired

result follows.

6. Assuming that $u(r,z) = R(r)Z(z)$ it follows from Problem 5

that $R = c_1 J_0(\lambda r) + c_2 Y_0(\lambda r)$ and $Z = k_1 e^{-\lambda z} + k_2 e^{\lambda z}$.

Since $u(r,z)$ is bounded as $r \to 0$ and $z \to \infty$ we require

that $c_2 = 0$, $k_2 = 0$. The B.C. $u(1,z) = 0$ requires that

$J_0(\lambda) = 0$ leading to an infinite set of discrete positive

eigenvalues $\lambda_1, \lambda_2, \ldots, \lambda_n, \ldots$. The fundamental solutions

of the problem are then $u_n(r,z) = J_0(\lambda_n r)e^{-\lambda_n z}$, $n =$

1,2,... . Thus assume $u(r,z) = \sum_{n=1}^{\infty} c_n J_0(\lambda_n r)e^{-\lambda_n z}$. The

B.C. $u(r,0) = f(r)$, $0 \leq r \leq 1$ requires that $u(r,0) =$

$\sum_{n=1}^{\infty} c_n J_0(\lambda_n r) = f(r)$ so

$c_n = \int_0^1 r\, f(r)J_0(\lambda_n r)dr / \int_0^1 r\, J_0^2(\lambda_n r)dr$, $n = 1,2,\ldots$.

7b. Again Θ periodic of period 2π implies $\lambda^2 = n^2$. Thus the

solutions to the D.E. are $R(r) = c_1 J_n(kr) + c_2 Y_n(kr)$ and

$\Theta(\theta) = d_1 \cos n\theta + d_2 \sin n\theta$.

9a. Substituting $u(\rho,\theta,\phi) = P(\rho)\Theta(\theta)\Phi(\phi)$ into Laplace's

equation leads to $\rho^2 P''/P + 2\rho P'/P = -(\csc^2\phi)\Theta''/\Theta - \Phi''/\Phi -$

$(\cot\phi)\Phi'/\Phi = \sigma$. In order to satisfy arbitrary

B.C. it can be shown that σ must be positive, so assume

$\sigma = \mu^2$. Thus $\rho^2 P'' + 2\rho P' - \mu^2 P = 0$. Then we have

$(\sin^2\phi)\Phi''/\Phi + (\sin\phi \cos\phi)\Phi'/\Phi + \mu^2 \sin^2\phi = -\Theta''/\Theta =$

α . The periodicity condition $\Theta(0) = \Theta(2\pi)$ requires that

$\sqrt{\alpha}$ be an integer λ so $\alpha = \lambda^2$. Hence $\Theta'' + \lambda^2\Theta = 0$ and

$(\sin^2\phi)\Phi'' + (\sin\phi \cos\phi)\Phi' + (\mu^2 \sin^2\phi - \lambda^2) = 0$.

10. The general solution to the Euler equation is $P = c_1 \rho^{r_1}$

$+ c_2 \rho^{r_2}$ where $r_1 = (-1+\sqrt{1+4\mu^2})/2 > 0$ and $r_2 =$

$(-1-\sqrt{1+4\mu^2})/2 < 0$. Since we want u to be bounded as $\rho \to 0$,

we set $c_2 = 0$. As found in Problem 10 of Section 4.2.1,

the solutions of Legendre's equation are either singular

at 1, at -1, or at both unless $\mu^2 = n(n+1)$, where n is an

integer. In this case, one of the two linearly
independent solutions is a polynomial denoted by P_n
(Problems 11 and 12 of Section 4.2.1). Since $r_1 =$
$(-1 + \sqrt{1+4n(n+1)})/2 = n$, the fundamental solutions of
this problem satisfying the finiteness condition are
$u_n(\rho,\phi) = \rho^n P_n(\cos \phi)$, $n = 1,2,\ldots$. It can be shown
that an arbitrary piecewise continuous function on $[-1,1]$
can be expressed as a linear combination of Legendre
polynomials. Hence we assume that $u(\rho,\phi) =$
$\sum_{n=1}^{\infty} c_n \rho^n P_n(\cos \phi)$. The B.C. $u(1,\phi) = f(\phi)$ requires that

$$u(1,\phi) = \sum_{n=1}^{\infty} c_n P_n(\cos \phi) = f(\phi), \quad 0 \le \phi \le \pi.$$ From Problem

15 of Section 4.2.1 we know that the $P_n(x)$ are
orthogonal. However here we have $P_n(\cos \phi)$ and thus we
must rewrite the equation in 9b to find $-[(\sin \phi)\Phi']' =$
$\mu^2(\sin\phi) \Phi$. Thus $P_n(\cos \phi)$ and $P_m(\cos \phi)$ are orthogonal
with weight function $\sin \phi$. Thus we must multiply the
series expansion for $f(\phi)$ by $\sin\phi \; P_m(\cos \phi)$ and
integrate from 0 to π to obtain
$$c_m = \int_0^{\pi} f(\phi)\sin \phi \; P_m(\cos \phi)d\phi / \int_0^{\pi} \sin\phi \; P_m^2(\cos \phi)d\phi.$$ To
obtain the answer as given in the text let $s = \cos \phi$.

Section 11.7, Page 651

1a. Rewrite $S_n(x)$ as $n\sqrt{x}/e^{nx^2/2}$ and use L'Hopitals Rule.

2b. Since $f(x)$ is defined only on the open interval we see that f(x) can get as close to 1 as desired, but f(x) is never equal to 1. Thus the least upper bound is 1, but there is no maximum value.

3. Expanding the integrand we get

$$R_n = \int_0^1 r(x)[f(x) - S_n(x)]^2 \, dx = \int_0^1 r(x)f^2(x)dx$$

$$- 2 \sum_{i=1}^n c_i \int_0^1 r(x)f(x)\phi_i(x)dx + \sum_{i=1}^n \sum_{j=1}^n c_i c_j \int_0^1 r(x)\phi_i(x)\phi_j(x)dx,$$

where the last term is obtained by calculating $S_n^2(x)$.

Using Eqs. (1) and (9) this becomes $R_n = \int_0^1 r(x)f^2(x)dx$

$$- 2 \sum_{i=1}^n c_i a_i + \sum_{i=1}^n c_i^2 = \int_0^1 r(x)f^2(x)dx - \sum_{i=1}^n a_i^2 + \sum_{i=1}^n (c_i -$$

$a_i)^2$, by completing the square. Since all terms involve a real quantity squared (and $r(x) > 0$) we may conclude R_n is minimized by choosing $c_i = a_i$. This can also be shown by calculating $\partial R_n/\partial c_i = 2(c_i - a_i)$ and setting equal to zero.

5b. From part a we have $f_0(x) = 1$ and thus $f_1(x) = c_1 + c_2 x$

must satisfy $(f_0, f_1) = \int_0^1 (c_1 + c_2 x)dx = 0$ and $(f_1, f_1) =$

$\int_0^1 (c_1 + c_2 x)^2 dx = 1$. Evaluating the integrals yields c_1

$+ c_2/2 = 0$ and $c_1^2 + c_1 c_2 + c_2^2/3 = 1$, which have the solution $c_1 = \sqrt{3}$, $c_2 = -2\sqrt{3}$ and thus $f_1(x) = \sqrt{3}(1-2x)$.

5c. $f_2(x) = c_1 + c_2x + c_3x^2$ must satisfy $(f_0,f_2) = 0$, (f_1,f_2)
$= 0$ and $(f_2,f_2) = 1$.

5d. For $g_2(x) = c_1 + c_2x + c_3x^2$ we have $(g_0,g_2) = 0$ and
$(g_1,g_2) = 0$, which yield the same ratio of coefficients
as found in 5c. Thus $g_2(x) = cf_2(x)$, where c may now be
found from $g_2(1) = 1$.

6. This problem follows the pattern of Problem 5 except now
the limits on the orthogonality integral are from -1 to

1. That is $(P_i,P_j) = \int_{-1}^{1} P_i(x) \, P_j(x)dx = 0$, $i \neq j$. For

$i = 0$ and $j = 1$ we have $\int_{-1}^{1} (c_1+c_2x)dx = (c_1x+c_2x^2/2)\Big|_{-1}^{1} =$

$2c_1 = 0$ and thus $P(1) = 1$ yields $P_1(x) = x$. The others
follow in a similar fashion.

7a. This part has essentially been worked in Problem 3 by
setting $c_i = a_i$.

7b. Eq.(6) shows that $R_n \geq 0$ since $r(x) \geq 0$ and thus

$\int_0^1 r(x)f^2(x)dx - \sum_{i=1}^{n} a_i^2 \geq 0$. The result follows.

7c. Since f is square integrable, $\int_0^1 r(x)f^2(x)dx = M < \infty$ and

therefore the monotone increasing sequence of partial

sums $T_n = \sum\limits_{i=1}^{n} a_i^2$ is bounded above. Thus $\lim\limits_{n \to \infty} T_n$ exists,

which proves the convergence of the given sum.

7e. By definition if $\sum\limits_{i=1}^{\infty} a_i \phi_i(x)$ converges to $f(x)$ in the

mean, then $R_n \to 0$ as $n \to \infty$. Hence $\int_0^1 r(x)f^2(x)dx = \sum\limits_{i=1}^{\infty} a_i^2$.

Conversely, if $\int_0^1 r(x)f^2(x)dx = \sum\limits_{i=1}^{\infty} a_i^2$, $\lim\limits_{n \to \infty} R_n = 0$ and

$\sum\limits_{i=1}^{\infty} a_i \phi_i(x)$ converges to $f(x)$ in the mean.

8. Bessel's inequality implies that $\sum\limits_{i=1}^{\infty} a_i^2$ converges and thus

the nth term $a_n \to 0$ as $n \to \infty$.

10. If the series were the eigenfunction series for a square

integrable function, the series $\sum\limits_{i=1}^{\infty} a_i^2$ would have to

converge. But $a_0 = 1$, $a_1 = 1/\sqrt{2}, \ldots$, $a_n = 1/\sqrt{n}, \ldots$,

and $\sum\limits_{n=1}^{\infty} a_n^2 = \sum\limits_{n=1}^{\infty} 1/n$ is the well-known harmonic series

which does not converge.